MATHEMATICAL METHODS IN ELECTRICAL ENGINEERING

Mathematical methods in electrical engineering

THOMAS B. A. SENIOR

Department of Electrical Engineering and Computer Science
University of Michigan

CAMBRIDGE UNIVERSITY PRESS
Cambridge, New York, Melbourne, Madrid, Cape Town, Singapore, São Paulo

Cambridge University Press
The Edinburgh Building, Cambridge CB2 8RU, UK

Published in the United States of America by Cambridge University Press, New York

www.cambridge.org
Information on this title: www.cambridge.org/9780521306614

First published 1986
Reprinted 1987, 1988, 1992
This digitally printed version 2008

A catalogue record for this publication is available from the British Library

Library of Congress Cataloguing in Publication data

Senior, Thomas B. A.
Mathematical methods in electrical engineering.
Includes index.
1. Electric engineering – Mathematics.
I. Title.
TK153.S46 1986 515 85-13275

ISBN 978-0-521-30661-4 hardback
ISBN 978-0-521-05677-9 paperback

Contents

Preface

For the past decade and more, all students in electrical and computer engineering at the University of Michigan have been required to take a course concerned with the mathematical methods for the solution of linear-systems problems. The course is typically taken in the junior year after completion of a basic four-semester mathematics sequence covering analytic geometry, matrices and determinants, differential and integral calculus, elementary differential equations, and so forth. Because it is the last course in mathematics that all must take for a bachelor's degree, a number of topics are included in addition to those customary in an introductory systems course, for example, functions of a complex variable with particular reference to integration in the complex plane, and Fourier series and transforms.

Some of the courses that make use of this material can be taken concurrently, thereby constraining the order in which the topics are covered. It is, for example, necessary that Laplace transforms be introduced early and the treatment carried to such a stage that the student is able to use the transform to solve initial-value problems. Fourier series must also be covered in the first half of the course, and the net result is an ordering that is different from the mathematically natural one, but that is quite advantageous in practice. Thus, the relatively simple material dealing with the Laplace transform and its applications comes at the beginning and provides the student with a sense of achievement prior to the introduction of the more abstract material on functions of a complex variable.

Although the mathematics is standard, we have not found a text that is suitable for a course like this one. Most books on linear systems cover

Laplace transforms and systems applications to the depth required and more, but do not include complex-variable theory or, possibly, Fourier transforms; and the well-established texts on applied mathematics do not present the material on linear systems that is vital to the course. For many years we have operated with course notes of varying degrees of completeness, and this text is an outgrowth of one version. Apart from the inclusion of some additional material on multivalued functions and Fourier transforms, the text conforms to the course as presently taught, and can be covered in one semester.

I am indebted to the many hundreds of students who have taken the course and whose comments, polite as well as pungent, have helped improve the presentation. It is also a pleasure to acknowledge the assistance of a former student, D. A. Ksienski, who contributed the section on the discrete Fourier transform, and Wanita Rasey, who typed the manuscript. Her speed and accuracy contributed to the rapid completion of the project.

THOMAS B. A. SENIOR

CHAPTER 1

Complex numbers

The need for an extension of the concept of real numbers became evident as early as the sixteenth century in connection with the solution of algebraic equations such as quadratics and cubics. At a time when even negative real numbers were widely regarded as absurd or fictitious, it is not surprising that the numbers involving $\sqrt{-1}$ introduced by Raphael Bombelli in 1572 did not meet with universal acceptance. Over two hundred years were to elapse before the so-called "imaginaries" were given a rigorous mathematical foundation through the work of Gauss, Cauchy, and others, leading to the theory of complex numbers we know today.

1.1 Definition

Any extension of real numbers is created by the definition of arithmetical operations involving them. As defined by the British mathematician William Rowan Hamilton (1805–65) in 1835, a complex number z is an ordered pair of real numbers x, y:

$$z = (x, y), \tag{1.1}$$

where x is called the real part of z, abbreviated as $x = \operatorname{Re} z$, and y is the imaginary part of z, abbreviated as $y = \operatorname{Im} z$.

If $z = (x, y)$ and $z' = (x', y')$ are two such numbers, then

(i) $z = z'$ if and only if $x = x'$ and $y = y'$;
(ii) $z + z' = (x + x', y + y')$;
(iii) $-z = (-x, -y)$, implying $z - z' = (x - x', y - y')$;
(iv) $zz' = (xx' - yy', xy' + yx')$.

It is now easy to show that the commutative and associative laws of addition and multiplication are satisfied, as well as the distributive law. Thus complex numbers obey the same fundamental laws of algebra as real numbers, and the algebra is therefore identical in form, though not in meaning, with the algebra of real numbers. Also

(v) if $z' = 1/z$ implying $zz' = 1$, it follows from (iv) that

$$z' = \left(\frac{x}{x^2 + y^2}, -\frac{y}{x^2 + y^2} \right)$$

provided $x^2 + y^2 \neq 0$.

It is customary to denote a complex number (x, y) whose imaginary part y is zero by the real number x. With this understanding, $x = (x, 0)$ and $y = (y, 0)$, and real numbers are merely a special case of the complex numbers defined above.

The more standard notation for a complex number is obtained by writing the particular complex number $(0, 1)$ as i (from the initial letter of the word imaginary). Then, from (iv),

$$i^2 = (0,1) \cdot (0,1) = (-1,0) = -1,$$

so that i may be regarded as the square root of the real number -1. The symbol i for $\sqrt{-1}$ was first introduced by the Swiss mathematician Leonhard Euler (1707–83) in a 1777 memoir that was not published until 1794. Its use did not receive wide acceptance until the 1847 publication of the classic work of Augustin-Louis Cauchy (1789–1857) on the theory of imaginaries. In terms of i

$$\begin{aligned} x + iy &= (x,0) + (0,1) \cdot (y,0) \\ &= (x,0) + (0 \cdot y - 1 \cdot 0, 0 \cdot 0 + 1 \cdot y) \\ &= (x,0) + (0, y) = (x, y) \end{aligned}$$

implying

$$z = x + iy. \tag{1.2}$$

By virtue of this, we see that in any operation involving sums and products it is permissible to treat x, y, and i as though they were real numbers with the proviso that i^2 must always be replaced by -1. In contrast to real numbers, however, complex numbers cannot be ordered, and a statement that $z > z'$ has no meaning.

1.2 Complex conjugate and modulus

A complex number closely related to $z = x + iy$ is the *complex conjugate* of z, denoted by $\bar{z}$, obtained by changing the sign of the imaginary part:

$$\bar{z} = (x, -y) = x - iy. \tag{1.3}$$

In effect, the "sign of i" has been changed. From Eqs. (1.1) and (1.3)

$$x = \frac{1}{2}(z + \bar{z}), \qquad y = \frac{1}{2i}(z - \bar{z}) \tag{1.4}$$

and also

$$z\bar{z} = x^2 + y^2.$$

The *modulus* of z, written as $|z|$ and often spoken as mod z, is defined as

$$|z| = \sqrt{x^2 + y^2}, \tag{1.5}$$

where the positive branch of the square root is implied. Thus, $|z|$ is a real nonnegative number and $|z| = 0$ if and only if $x = y = 0$. Clearly

$$|z| = |\bar{z}| = \sqrt{z\bar{z}} \tag{1.6}$$

and in the case of a real number x

$$|x| = \sqrt{x^2} = \begin{cases} x & \text{if } x \geq 0, \\ -x & \text{if } x < 0. \end{cases}$$

In any mathematical proof involving $|z|$ it is necessary to interpret the modulus using Eq. (1.5) or (1.6). As an example we have

$$|zz'|^2 = zz'\overline{zz'} = z\bar{z}z'\bar{z}' = |z|^2|z'|^2,$$

from which it follows that

$$|zz'| = |z|\,|z'|.$$

There is an analogous result for a product of three (or more) numbers.

In contrast to complex numbers themselves, their moduli (being real) can be ordered and inequalities established among them. Thus,

$$|z| = \sqrt{x^2 + y^2} \geq x, y,$$

that is,

$$|z| \geq \operatorname{Re} z, \operatorname{Im} z. \tag{1.7}$$

Also

$$|z + z'|^2 = (z + z')(\bar{z} + \bar{z}')$$
$$= z\bar{z} + \bar{z}z' + z\bar{z}' + z'\bar{z}'$$
$$= |z|^2 + 2\,\mathrm{Re}\,\bar{z}z' + |z'|^2$$
$$\leq |z|^2 + 2|\bar{z}z'| + |z'|^2$$
$$= |z|^2 + 2|z|\,|z'| + |z'|^2$$
$$= (|z| + |z'|)^2$$

and hence

$$|z + z'| \leq |z| + |z'|, \tag{1.8}$$

a formula that is readily extended to any finite number of complex numbers. Observe the variety of properties that have been used in proving this result. Similarly,

$$|z - z'| \geq |(|z| - |z'|)|. \tag{1.9}$$

Equations (1.8) and (1.9) are the triangle inequalities, so named because of their geometrical interpretation.

Exercises

If $z = 1 + i$ and $z' = 3 - 4i$, find the real and imaginary parts of the following:

*1. $(z - z')^2$. 2. $\dfrac{z}{z'}$.

3. z^{-2}. 4. zz'.

5. Prove that $\mathrm{Re}\, i\bar{z} = \mathrm{Im}\, z$ and $\mathrm{Im}\, i\bar{z} = \mathrm{Re}\, z$.

6. Prove that z is real or purely imaginary if and only if $\bar{z}^2 = z^2$.

7. Prove that if the product of two complex numbers is zero, at least one of them must be zero.

Evaluate

*8. $\left|\dfrac{1 + 4i}{1 - 4i}\right|^2$. 9. $\left|\dfrac{1 - i}{2 + i}\right|^2$.

*Here and in the following sets of exercises, an asterisk indicates that the answer is listed at the end of the book.

10. $\left|\dfrac{\bar{z}}{z}\right|$. 11. $\left|\dfrac{z + i\bar{z} - 1}{z + i\bar{z} - i}\right|$.

12. Verify the triangle inequality (1.8) for $z = 1 + i$, $z' = -2 + 3i$.

13. Verify the triangle inequality (1.9) for $z = 3 - 4i$, $z' = 2 - i$.

14. Prove that $|z + z'|^2 + |z - z'|^2 = 2(|z|^2 + |z'|^2)$.

15. Using the result of Exercise 14, show

$$\left|z_1 + \sqrt{z_1^2 - z_2^2}\right| + \left|z_1 - \sqrt{z_1^2 - z_2^2}\right| = |z_1 + z_2| + |z_1 - z_2|.$$

Hint: Consider the square of the left-hand side.

16. Prove that $|z - z'| \geq |(|z| - |z'|)|$.

1.3 Geometric representation: The complex plane

Although the definition of z does not require that a complex number be represented (or even be capable of representation) geometrically, there are many advantages in doing so, and definition as an ordered pair of real numbers naturally suggests the interpretation of x and y as the Cartesian coordinates of a point in a plane. If the horizontal axis is taken to be the x (or real) axis and the vertical axis to be the y (or imaginary) axis as shown in Fig. 1.1, the resulting plane is called the *complex* (z) *plane*. A somewhat archaic term is the *Argand diagram*, after the French mathematician Jean Robert Argand (1768–1822), who published his graphic representation in 1806, nine years after a similar description by the Norwegian mathematician Caspar Wessel (1745–1818).

Any complex number z is represented by a point whose Cartesian coordinates are the real and imaginary parts of z, and conversely, each point in the plane represents a complex number. The complex-conjugate point $\bar{z}$ is the mirror image of z in the real axis. From Pythagoras's theorem the distance of the point z (or $\bar{z}$) from the origin O of the plane is $|z|$. Likewise $|z - z'|$ is the distance between the points z and z', and for given z' and a, the equation

$$|z - z'| = a$$

defines the set of points z lying on the circumference of a circle of radius a centered on the point z'. By considering the triangle formed by the points $z + z'$, z, and the origin, the triangle inequality (1.8) is seen to imply that the sum of the lengths of two sides of a triangle is greater than or equal to the length of the third side.

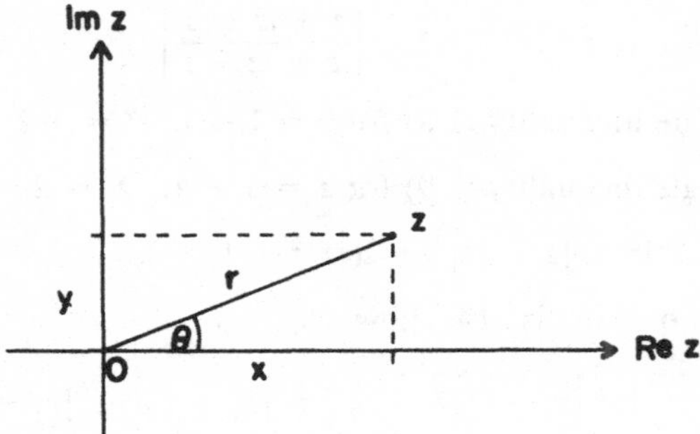

Fig. 1.1: The complex z plane.

Any point other than O can also be specified using polar coordinates (r, θ) where r is the distance from the origin and θ is the angle between the line Oz and the positive x axis. Clearly

$$r = |z| \tag{1.10}$$

and

$$x = r\cos\theta, \qquad y = r\sin\theta, \tag{1.11}$$

so that

$$z = x + iy = r(\cos\theta + i\sin\theta). \tag{1.12}$$

Euler's formula, published in 1749, is

$$e^{i\theta} = \cos\theta + i\sin\theta. \tag{1.13}$$

It serves also to define

$$\cos\theta = \frac{1}{2}(e^{i\theta} + e^{-i\theta}), \tag{1.14}$$

$$\sin\theta = \frac{1}{2i}(e^{i\theta} - e^{-i\theta}). \tag{1.15}$$

(Equation (1.14) was actually derived by Euler in 1740.) From Eqs. (1.10), (1.12), and (1.13) we have

$$z = |z|e^{i\theta}, \tag{1.16}$$

where $\theta = \arctan y/x$, and this gives a representation of z in terms of polar coordinates. The angle θ is called the *argument of* z, and written $\theta = \arg z$.

A coordinate system must uniquely define a point, but it is not always true that a point uniquely defines a coordinate. An example is provided by the angle $\theta = \arg z$. This is not unique since, to any one value $\theta^{(1)}$, we can add integer multiples of 2π without affecting the location of the point z. Thus, $\arg z$ is a multi-valued function of z and, in general,

$$\theta = \theta^{(1)} + 2m\pi, \qquad m = 0, \pm 1, \pm 2, \ldots . \tag{1.17}$$

It is sometimes convenient to restrict the allowed range of $\arg z$ by defining a principal value, denoted by $\operatorname{Arg} z$, having $-\pi < \operatorname{Arg} z \leq \pi$. The result is a single-valued function of z, but this has been achieved at the expense of making $\operatorname{Arg} z$ discontinuous as the point z crosses the negative real axis.

The polar form also suggests an operational interpretation of the complex number i. From Eq. (1.16)

$$iz = |z|e^{i(\theta+\pi/2)},$$

showing that multiplication by i has the effect of an anticlockwise rotation through an angle $\pi/2$. Since two such rotations are equivalent to multiplication by -1, this interpretation is consistent with the fact that $i = \sqrt{-1}$.

1.4 Powers and roots

The polar representation of complex numbers is particularly useful for multiplication and division. If, for example,

$$z = z_1 z_2,$$

where

$$z_1 = r_1 e^{i\theta_1}, \qquad z_2 = r_2 e^{i\theta_2},$$

it follows immediately that

$$z = r_1 r_2 e^{i(\theta_1+\theta_2)}$$

implying

$$|z| = r_1 r_2 = |z_1|\,|z_2|$$

and

$$\arg z = \theta_1 + \theta_2 = \arg z_1 + \arg z_2.$$

Although $\arg z$ is, of course, not unique,

$$x = \operatorname{Re} z = |z_1|\,|z_2|\cos(\theta_1 + \theta_2),$$
$$y = \operatorname{Im} z = |z_1|\,|z_2|\sin(\theta_1 + \theta_2)$$

are unique, since the addition of an integer multiple of 2π to $\theta_1 + \theta_2$ does not affect the cosine or sine functions. Similarly, if

$$z = \frac{z_1}{z_2}$$

then

$$|z| = \left|\frac{z_1}{z_2}\right| \quad \text{and} \quad \arg z = \theta_1 - \theta_2,$$

from which we have

$$x = \operatorname{Re} z = \left|\frac{z_1}{z_2}\right| \cos(\theta_1 - \theta_2),$$

$$y = \operatorname{Im} z = \left|\frac{z_1}{z_2}\right| \sin(\theta_1 - \theta_2).$$

It is a simple matter to extend the multiplication formula to a product of n complex numbers $z_1, z_2, \ldots, z_n$, and in the particular case in which these are identical and equal to z,

$$z^n = (re^{i\theta})^n = r^n e^{in\theta}, \tag{1.18}$$

where $r = |z|$ and $\theta = \arg z$. Hence

$$\operatorname{Re} z^n = |z|^n \cos n\theta,$$

$$\operatorname{Im} z^n = |z|^n \sin n\theta,$$

and these are unique for any integer n, positive or negative. Moreover, from Eq. (1.18) with $r = 1$,

$$(\cos\theta + i\sin\theta)^n = \cos n\theta + i \sin n\theta, \tag{1.19}$$

a formula that is usually credited to the French mathematician Abraham de Moivre (1667–1754).

Whereas z^n is a single-valued function of z, $z^{1/n}$ is a multivalued function. This is evident if we write $\xi = z^{1/n}$. Then

$$\xi^n - z = 0$$

and the determination of ξ is equivalent to finding the n roots of this equation (i.e., the n roots of z).

The multiplicity of values comes about because of the nonuniqueness of $\theta = \arg z$. If $\theta^{(1)}$ is a particular value, then

$$z = re^{i\theta} = re^{i(\theta^{(1)} + 2m\pi)}$$

for any integer m, and hence

$$z^{1/n} = r^{1/n} e^{i(\theta^{(1)} + 2m\pi)/n} \tag{1.20}$$

with $r = |z|$. Separation into real and imaginary parts now gives

$$\operatorname{Re} z^{1/n} = r^{1/n} \cos\frac{1}{n}(\theta^{(1)} + 2m\pi),$$

$$\operatorname{Im} z^{1/n} = r^{1/n} \sin\frac{1}{n}(\theta^{(1)} + 2m\pi)$$

for any integer m, and by assigning m the n consecutive values $0, 1, 2, \ldots, n-1$ we generate n distinct values of $z^{1/n}$. Other m merely reproduce the complex numbers already found, and thus

$$z^{1/n} = |z|^{1/n}\left\{\cos\frac{1}{n}(\theta^{(1)} + 2m\pi) + i\sin\frac{1}{n}(\theta^{(1)} + 2m\pi)\right\}, \tag{1.21}$$

where $\theta^{(1)} = \arctan y/x$ and $m = 0, 1, 2, \ldots, n-1$.

As evident from this, the n values of $z^{1/n}$ (or the n roots of z) all have the same moduli and lie on a circle of radius $|z|^{1/n}$, whose center is the origin of the complex plane. They are equally spaced around the circle, and the arguments of two adjacent roots differ by $2\pi/m$. Once $\theta^{(1)}$ and $|z|^{1/n}$ are computed, it is therefore a simple matter to locate the roots in the complex plane.

If p and q are integers not equal to zero, the same reasoning as above shows that $z^{p/q}$ has q distinct values, which are given by

$$z^{p/q} = |z|^{p/q}\left\{\cos\frac{p}{q}(\theta^{(1)} + 2m\pi) + i\sin\frac{p}{q}(\theta^{(1)} + 2m\pi)\right\} \tag{1.22}$$

with $m = 0, 1, 2, \ldots, q-1$.

Example 1. Find the real and imaginary parts of the cube roots of unity.

The task is to determine $z^{1/3}$ with $z = 1$. Since $x = 1$ and $y = 0$, $|z| = 1$ and the natural choice of $\theta^{(1)}$ is zero. Hence, from Eq. (1.21),

$$1^{1/3} = \cos\frac{2m\pi}{3} + i\sin\frac{2m\pi}{3}, \qquad m = 0, 1, 2.$$

On giving m its three consecutive values, we obtain

$$1, \; -\frac{1}{2} + i\frac{\sqrt{3}}{2}, \; -\frac{1}{2} - i\frac{\sqrt{3}}{2}$$

and these are shown in Fig. 1.2. We observe that the roots are 1, ω, and ω^2, where $\omega = \frac{1}{2}(-1 + i\sqrt{3})$.

Example 2. Find the values of $(8i)^{1/3}$ and display them in the complex plane.

For $z = 8i$ we have $|z| = 8$ and the natural choice of $\theta^{(1)}$ is $\pi/2$. Hence

$$\begin{aligned}(8i)^{1/3} &= 8^{1/3}e^{i(1/2+2m)\pi/3}\\ &= 2e^{i\pi/6+2im\pi/3}, \qquad m = 0, 1, 2.\end{aligned}$$

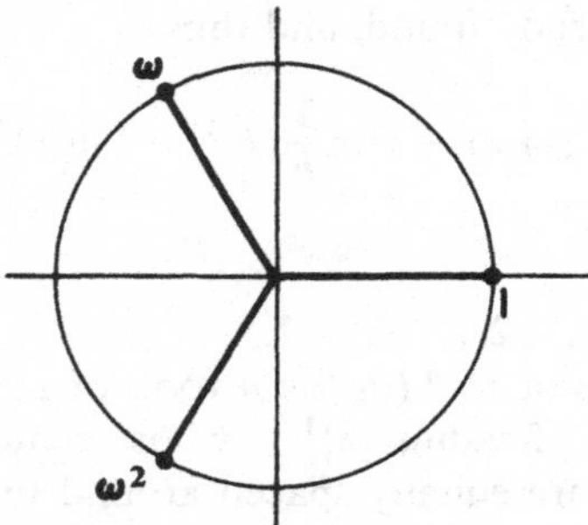

Fig. 1.2: Cube roots of 1.

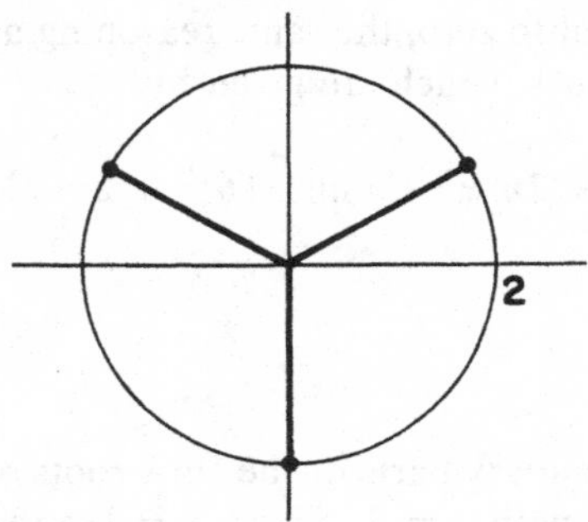

Fig. 1.3: Cube roots of $8i$.

These all lie on a circle of radius 2 and are shown in Fig. 1.3.

Exercises

If $z_1 = \frac{1}{2}(\sqrt{3} - i)$ and $z_2 = \frac{1}{2}(1 + i)$, use the polar form to compute the real and imaginary parts of the following:

*1. $\dfrac{z_1^3}{z_2}$.

2. $(z_1 z_2)^4$.

3. $\left(\dfrac{i - z_1}{i - z_2}\right)^3$.

4. $\dfrac{(1 - z_2)^5}{z_1^3}$.

Solve the following equations and display the roots in the complex plane:

5. $z^4 + 81 = 0$.

6. $z^6 + 1 = \sqrt{3}\, i$.

Find the real and imaginary parts of $z^{1/2}$, where

*7. $z = 5 - 12i$.

8. $z = 8 + 4\sqrt{5}\, i$.

9. Find the real and imaginary parts of the roots of $z^4 + z^2 + 1 = 0$.

10. Show that the nth roots of z can be written as

$$z_1, \omega z_1, \omega^2 z_1, \ldots, \omega^{n-1} z_1,$$

where z_1 is a particular root, and $1, \omega, \omega^2, \ldots, \omega^{n-1}$ are the nth roots of unity.

*11. Determine the set of points (x, y) defined by $|z + 2| = 2|z - 1|$.

12. Physically describe and analytically determine the region in the complex plane defined by

$$|z - a| + |z - b| < k,$$

where a, b, and k are real with $0 < a - b < k$.

1.5 Exponential and logarithmic functions

A general treatment of functions of a complex variable is reserved for Chapter 5, but there are two special functions whose properties are needed before then. They are the exponential and its inverse, the logarithm.

The exponential function e^z is defined by its power series

$$e^z = 1 + z + \frac{z^2}{2!} + \frac{z^3}{3!} + \cdots = \sum_{n=0}^{\infty} \frac{z^n}{n!}, \tag{1.23}$$

which is absolutely convergent for all z. In the particular case in which $z = x$, the power series becomes that defining the real exponential e^x. By multiplication of power series it can be shown that

$$e^z \cdot e^{z'} = e^{z+z'}$$

and as a special case

$$e^z = e^{x+iy} = e^x e^{iy},$$

that is,

$$e^z = e^x(\cos y + i \sin y)$$

using Euler's formula (1.13). Since $|e^{iy}| = \sqrt{\cos^2 y + \sin^2 y} = 1$,

$$|e^z| = |e^x|\,|e^{iy}| = e^x.$$

Also, using term-by-term differentiation of the power series in Eq. (1.23),

$$\frac{d}{dz} e^z = e^z.$$

When x is positive, the equation $e^u = x$ has one real solution $u = \ln x$ where ln denotes the natural logarithm (base e); if z is complex and nonzero, the corresponding equation $e^w = z$ has an infinity of solutions, each of which is called a logarithm of z. With $w = u + iv$ we have

$$e^w = e^{u+iv} = z.$$

Thus, v is one of the values of $\arg z$ and $e^u = |z|$, implying $u = \ln|z|$. It now follows that every solution of the equation $e^w = z$ has the form

$$w = \ln|z| + i \arg z,$$

and since $\arg z$ has an infinity of values, so does the logarithm of the complex number z, with pairs differing by $2\pi i$. We write

$$\log z = \ln|z| + i \arg z, \tag{1.24}$$

where $\log z$ is an infinitely many-valued function of z.

The *principal value* of $\log z$ is obtained by giving $\arg z$ its principal value $\operatorname{Arg} z$, $-\pi < \operatorname{Arg} z \leq \pi$, and since it is identical to the ordinary logarithm when z is real and positive, it will be denoted by $\ln z$.

A function closely related to the logarithm is the general power ξ^z. If ξ is real and positive and equal to a, the function is unambiguously defined as

$$a^z = e^{z \ln a}.$$

For complex $\xi \neq 0$, the principal value of ξ^z is the number uniquely determined by the equation

$$\xi^z = e^{z \ln \xi},$$

where $\ln \xi$ is, as above, the principal value of $\log \xi$. By choosing other values of $\log \xi$, we obtain the subsidiary values as well, and all are contained in the formula

$$\xi^z = e^{z(\ln \xi + 2m\pi i)} \tag{1.25}$$

with $m = 0, \pm 1, \pm 2, \dots$. Hence, ξ^z has an infinity of values, but one (and only one) principal value corresponding to $m = 0$.

Example. In a 1746 letter Euler claimed that $i^i = 0.207\,879\,5763$. Show that this is one of the infinity of possible real numbers and obtain the general expression.

From Eq. (1.25) with $z = i$ and $\ln i = \ln 1 + i\pi/2 = i\pi/2$, we have

$$i^i = e^{i(i\pi/2 + 2m\pi i)} = e^{-\pi/2 - 2m\pi}$$

with $m = 0, \pm 1, \pm 2, \dots$. The principal value ($m = 0$) is

$$e^{-\pi/2} = 0.207\,879\,5763\dots.$$

CHAPTER 2

Laplace transforms

The Laplace transform is a key part of the operational calculus that is at the heart of the modern approach to systems theory. Its application was stimulated by the work of Oliver Heaviside (1850–1925), an English physicist, who developed methods for the systematic solution of ordinary differential equations. His methods were largely intuitive and without proof, but in the years since, the mathematical foundations have been rigorously established, initially for the so-called p and D methods, and most recently in the context of the Laplace transform (after Pierre Simon de Laplace (1749–1827), the French mathematician).

2.1 Definition of the Laplace transform

Let $f(t)$ be a real or complex function of the real variable t. We shall think of t as denoting time (in which case $-\infty < t < \infty$), but it could also represent, for example, a spatial variable. The Laplace transform of $f(t)$ is then

$$\int_0^\infty e^{-st} f(t)\, dt,$$

where s is some parameter, in general complex. The result is clearly a function of s, which is obtained by applying the integral operator

$$\int_0^\infty e^{-st}(\cdots)\, dt$$

to $f(t)$, and to emphasize the operational aspect, we shall write

$$\mathscr{L}\{f(t)\} = \int_0^\infty e^{-st} f(t)\,dt = F(s). \tag{2.1}$$

This is the definition of the (unilateral or one-sided) Laplace transform, and of what we mean by the operator $\mathscr{L}$.

Notation: We shall use lowercase letters to denote functions of t and reserve the corresponding uppercase letters to denote the transforms, which are functions of s. When it is necessary to display the real and imaginary parts of s, we shall write

$$s = \sigma + i\omega,$$

where $\sigma = \operatorname{Re} s$ and $\omega = \operatorname{Im} s$.

The Laplace transform is just one example of a class of integral transforms (or operators) that map functions of one variable (in this case t) into functions of another variable. We will meet another example later in this book, and it is possible to treat the Laplace transform as part of a general theory of integral operators by defining the class of functions to which they are applicable, their mapping properties, and so forth. But to do so would risk making a rather elementary operator seem quite complicated. Instead, therefore, we shall consider the Laplace transform alone, and build up an understanding of its properties and applications bit by bit.

Let us first see what happens when we apply the transform to a few simple functions. This will bring out a requirement of all infinite integrals (as the transform is), namely, that the integral must converge. Just as an infinite series is meaningless unless it converges (i.e., has a finite sum), so it is with an infinite integral. By definition

$$\int_0^\infty g(t)\,dt = \lim_{T\to\infty} \int_0^T g(t)\,dt$$

and the infinite integral converges only if the limit on the right-hand side is finite and unique.

Example 1. Determine $\mathscr{L}\{1\}$.

From Eq. (2.1) with $f(t) = 1$,

$$\mathscr{L}\{1\} = \int_0^\infty e^{-st}\,dt = \left[-\frac{1}{s}e^{-st}\right]_0^\infty.$$

Note that the limits are on t. On inserting them

$$\mathscr{L}\{1\} = \frac{1}{s} - \frac{1}{s}\lim_{t\to\infty} e^{-st}. \tag{2.2}$$

Since

$$e^{-st} = e^{-(\sigma+i\omega)t} = e^{-\sigma t}e^{-i\omega t},$$

it follows that

$$|e^{-st}| = |e^{-\sigma t}|\,|e^{-i\omega t}| = e^{-\sigma t}.$$

If $\sigma < 0$, the limit in Eq. (2.2) is infinite and $\mathscr{L}\{1\}$ is meaningless, whereas if $\sigma > 0$, the limit is zero. If $\sigma = 0$, $e^{-st} = e^{-i\omega t}$, and though the *magnitude* of this is unity (and therefore finite) for all t, it does not have a unique limit as $t \to \infty$. Thus, the integral converges and the transform has a meaning only if $\sigma > 0$, in which case

$$\mathscr{L}\{1\} = \frac{1}{s}, \qquad \sigma > 0. \tag{2.3}$$

Example 2. Determine $\mathscr{L}\{e^{at}\}$ where a is a complex constant.
From Eq. (2.1) with $f(t) = e^{at}$,

$$\mathscr{L}\{e^{at}\} = \int_0^\infty e^{-st}e^{at}\,dt = \int_0^\infty e^{-(s-a)t}\,dt.$$

Since this is now the same as Example 1 with s replaced by $s - a$,

$$\mathscr{L}\{e^{at}\} = \frac{1}{s-a}, \qquad \sigma > \operatorname{Re} a. \tag{2.4}$$

Example 3. Determine $\mathscr{L}\{t\}$.
From Eq. (2.1) with $f(t) = t$,

$$\begin{aligned}\mathscr{L}\{t\} &= \int_0^\infty e^{-st}t\,dt \\ &= \left[-\frac{t}{s}e^{-st}\right]_0^\infty + \frac{1}{s}\int_0^\infty e^{-st}\,dt \qquad \text{by parts} \\ &= -\frac{1}{s}\lim_{t\to\infty} te^{-st} + \frac{1}{s}\mathscr{L}\{1\}\end{aligned}$$

since the term in square brackets is zero when $t = 0$. The limit exists only if $\sigma > 0$, which is also the condition for the existence of $\mathscr{L}\{1\}$. Hence, from the result of Example 1,

$$\mathscr{L}\{t\} = \frac{1}{s^2}, \qquad \sigma > 0. \tag{2.5}$$

Example 4. Determine $\mathscr{L}\{\sin t\}$.

From Eq. (2.1) with $f(t) = \sin t$,

$$F(s) = \mathscr{L}\{\sin t\} = \int_0^\infty e^{-st} \sin t \, dt$$

$$= [-e^{-st}\cos t]_0^\infty - \int_0^\infty se^{-st}\cos t \, dt \qquad \text{by parts}$$

$$= 1 - s\int_0^\infty e^{-st}\cos t \, dt$$

provided $\sigma > 0$ so that $\lim_{t\to\infty} e^{-st}\cos t = 0$. Similarly

$$\int_0^\infty e^{-st}\cos t \, dt = [e^{-st}\sin t]_0^\infty + \int_0^\infty se^{-st}\sin t \, dt \qquad \text{by parts}$$

$$= sF(S)$$

since $\sin t = 0$ when $t = 0$. Hence

$$F(s) = 1 - s^2F(s)$$

giving

$$F(s) = \mathscr{L}\{\sin t\} = \frac{1}{s^2+1}, \qquad \sigma > 0. \tag{2.6}$$

From these examples it is evident that the factor e^{-st} in the transform integral acts like a *convergence factor* in that the allowed values of σ are those which make the integral converge. Since only the real part of the complex-variable s is restricted, the corresponding region in the complex s plane is the half-plane to the right of the stated σ. With σ chosen in this manner, the contribution from the upper limit of integration is always zero.

It could be that the integral will converge even if σ is negative. In Example 2, if $a = -2$, the condition for the existence of the transform is $\sigma > -2$, but for reasons that will become evident later, it is desirable to confine σ to nonnegative values only. Thus

$$\mathscr{L}\{e^{-2t}\} = \frac{1}{s+2}, \qquad \sigma \geq 0.$$

As seen from the preceding examples, σ is now restricted to that part of the right-half complex s plane (with imaginary axis included) that lies to the right of the singularity of $F(s)$ with largest real part.

Exercises

For the following $f(t)$ where α is real ($\alpha \geq 0$), find $\mathscr{L}\{f(t)\}$ by evaluating the integral (2.1) and state the allowed values of σ.

*1. t^2.
2. te^{3t}.
3. $\cos t$.
4. $t \sin \alpha t$.
*5. $\cosh \alpha t$.
6. $\sinh \alpha t$.
7. $e^{-t}\cosh \alpha t$.
8. $\sinh(t - \alpha)$.

2.2 Piecewise-continuous functions

For a function $f(t)$ where t is a real variable, the concept of continuity is, it is hoped, a familiar one. If, for example, the function is real so that it is easily plotted, continuity implies that it can be drawn without lifting the pencil from the paper. More mathematically: $f(t)$ is continuous for $a \leq t \leq b$ if, for any t in this range,

$$\lim_{\Delta t \to 0} f(t + \Delta t) \quad \text{exists and is unique.}$$

The limit, if it exists, is $f(t)$, the value of the function at t.

In the preceding definition the uniqueness of the limit is important because Δt can tend to zero through positive values, in which case t is approached from the right, or through negative values (approached from the left). This leads to the idea of right- and left-hand limits:

$$\text{right-hand limit} = \lim_{\Delta t \to 0+} f(t + \Delta t) = f(t+),$$

$$\text{left-hand limit} = \lim_{\Delta t \to 0-} f(t + \Delta t) = f(t-).$$

If t is a point of continuity of $f(t)$, the right- and left-hand limits exist and are equal (i.e., $f(t+) = f(t-)$), which we write simply as $f(t)$.

In contrast, suppose the limits exist but are *not* equal. The function shown in Fig. 2.2 is discontinuous at the points $t = t_1$ and $t = t_2$, where the right- and left-hand limits exist but are not equal. At all other points in $a \leq t \leq b$, $f(t)$ is continuous. By definition:

> $f(t)$ is piecewise (or sectionally) continuous in $a \leq t \leq b$ if it is continuous except at a finite number of points $t = t_n$ where it has finite jump discontinuities.

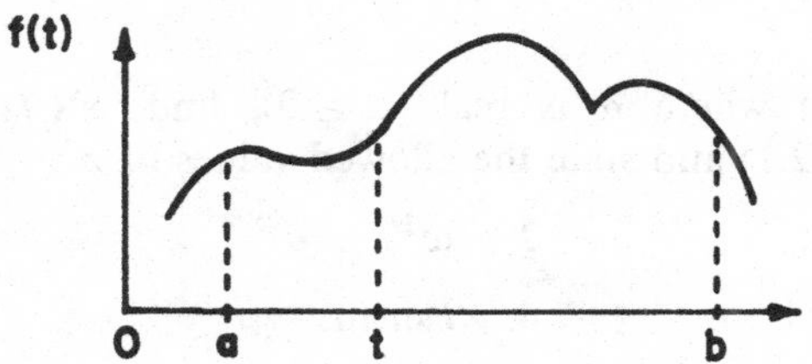

Fig. 2.1: Example of a function $f(t)$ continuous for $a \leq t \leq b$.

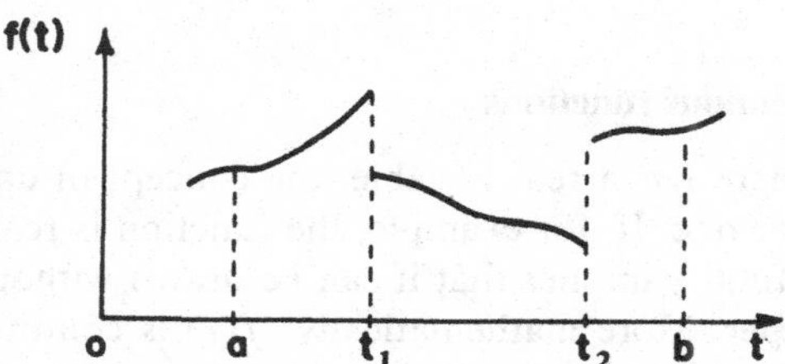

Fig. 2.2: Example of a function $f(t)$ which is piecewise-continuous in $a \leq t \leq b$.

There are several features of this definition that should be noted. In the first place, a piecewise-continuous function can have only a *finite* number of jump discontinuities in any *finite* range of t. At each such point, the right- and left-hand limits both exist, implying that the jump itself is finite. Thus, $\tan t$ is not a piecewise-continuous function in $0 \leq t \leq \pi$, since the discontinuity at $t = \pi/2$ is infinite. Finally we note that at a jump discontinuity $f(t)$ is undefined. Though it might seem logical to assign to $f(t)$ the average of its right- and left-hand limits, this is not implied by the definition, nor will it be assumed.

Unit-step function

The unit-step function $u(t)$ is defined as

$$u(t) = \begin{cases} 0 & \text{for } t < 0, \\ 2 & \text{for } t > 0. \end{cases} \tag{2.7}$$

Thus, $u(t)$ is piecewise-continuous in $-\infty < t < \infty$ with a single jump discontinuity at $t = 0$, where $u(t)$ is undefined. The function is plotted in Fig. 2.3.

In accordance with our convention, the lowercase letter is used for the function. Some other texts use U, and still others refer to it as the

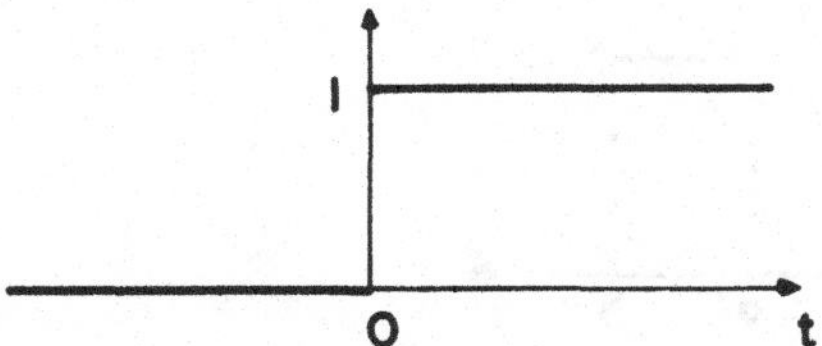

Fig. 2.3: The unit-step function $u(t)$.

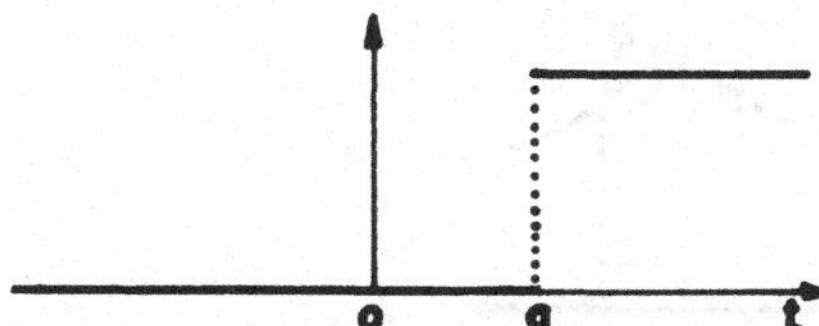

Fig. 2.4: Turn-on at $t = a$.

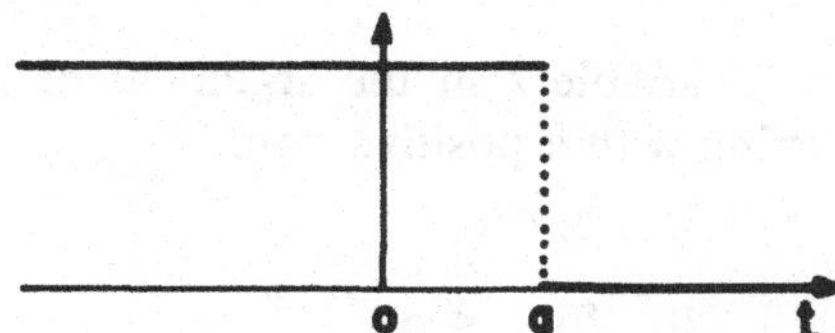

Fig. 2.5: Turn-off at $t = a$.

Heaviside function, after Oliver Heaviside, who first introduced it. $u(t)$ represents the turn-on action of an ideal switch. It is a very useful function, and we can do many things with it, but its mathematical meaning is neither more nor less than is indicated by Eq. (2.7). In particular, the only "action" occurs when the argument of the function is zero: For negative arguments the function is zero; for positive arguments the function is unity. Some examples of the unit-step function in use are as follows.

(i) Turn-on at $t = a$ (Fig. 2.4):

$$u(t-a) = \begin{cases} 0 & \text{for } t < a, \\ 1 & \text{for } t > a. \end{cases}$$

(ii) Turn-off at $t = a$ (Fig. 2.5):

$$1 - u(t-a) = \begin{cases} 1 & \text{for } t < a, \\ 0 & \text{for } t > a. \end{cases}$$

The same operation is accomplished by $u(a - t)$, but it is

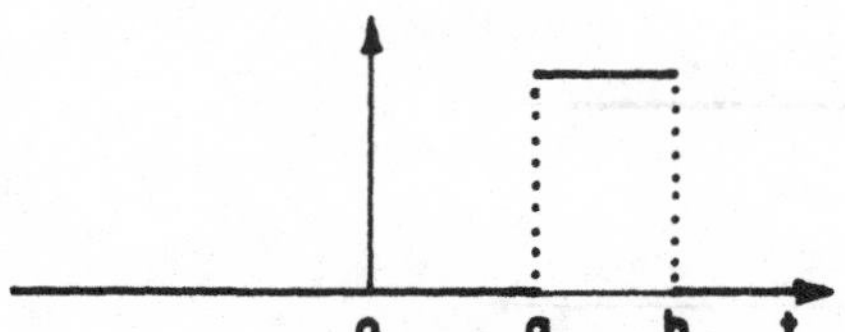

Fig. 2.6: Rectangular pulse.

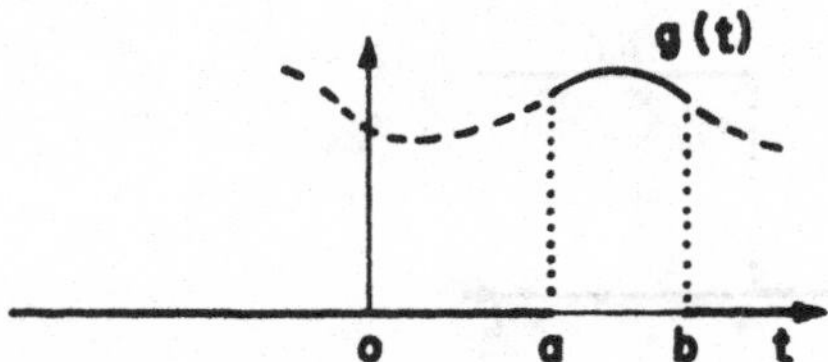

Fig. 2.7: Modulated pulse.

convenient to have the variable t in the argument of the step function always occurring with a positive sign.

(iii) Rectangular pulse (Fig. 2.6):

$$u(t-a) - u(t-b) = \begin{cases} 0 & \text{for } t < a, \\ 1 & \text{for } a < t < b, \\ 0 & \text{for } b < t, \end{cases}$$

where we have assumed that $a < b$.

(iv) Modulated pulse (Fig. 2.7):

$$g(t)\{u(t-a) - u(t-b)\} = \begin{cases} 0 & \text{for } t < a, \\ g(t) & \text{for } a < t < b, \\ 0 & \text{for } b < t, \end{cases}$$

where we have again assumed that $a < b$. The expression on the left is easily plotted and just as easily read; that is, we have the function $g(t)$ that is turned on at $t = a$ and off at $t = b$.

(v) Change of function: If

$$f(t) = \begin{cases} 1 & \text{for } t < a, \\ e^{-t} & \text{for } a < t, \end{cases}$$

then

$$f(t) = 1 + (e^{-t} - 1)u(t-a).$$

Example 1. Determine $\mathscr{L}\{1 - u(t-2)\}$.
Since

$$f(t) = \begin{cases} 1 & \text{for } t < 2, \\ 0 & \text{for } 2 < t, \end{cases}$$

we have

$$\mathscr{L}\{f(t)\} = \int_0^2 e^{-st}\,dt = \left[-\frac{e^{-st}}{s}\right]_0^2 = \frac{1}{s}(1 - e^{-2s}). \tag{2.8}$$

Strictly speaking, since the transform here is a finite integral, no restriction on σ is necessary.

Example 2. Determine $\mathscr{L}\{t^n u(t)\}$ where n is a nonnegative integer.
From Eq. (2.1) with $f(t) = t^n u(t)$,

$$\mathscr{L}\{t^n u(t)\} = \int_0^\infty e^{-st} t^n u(t)\,dt = \int_0^\infty e^{-st} t^n\,dt$$

and we observe that the step function $u(t)$ has no effect on the transform. If $n \geq 1$

$$\mathscr{L}\{t^n u(t)\} = \left[-\frac{t^n}{s} e^{-st}\right]_0^\infty + \frac{1}{s}\int_0^\infty e^{-st} n t^{n-1}\,dt \qquad \text{by parts}$$

$$= \frac{n}{s}\mathscr{L}\{t^{n-1}u(t)\}, \qquad \sigma > 0$$

and by repeating this process as often as necessary, we obtain

$$\mathscr{L}\{t^n u(t)\} = \frac{n}{s}\cdot\frac{n-1}{s}\cdot\frac{n-2}{s}\cdots\frac{1}{s}\mathscr{L}\{1\}.$$

But, from Eq. (2.3),

$$\mathscr{L}\{1\} = \frac{1}{s}, \qquad \sigma > 0.$$

Hence

$$\mathscr{L}\{t^n u(t)\} = \frac{n!}{s^{n+1}}, \qquad \sigma > 0. \tag{2.9}$$

Exercises

For the first two problems, express $f(t)$ in terms of unit-step functions.

*1. $f(t) = \begin{cases} 0 & \text{for } t < 0, \\ e^{-t^2} & \text{for } 0 < t < 2, \\ 0 & \text{for } 2 < t. \end{cases}$

2. $f(t) = \begin{cases} e^t & \text{for } t < 0, \\ t & \text{for } 0 < t < 1, \\ 1 & \text{for } 1 < t. \end{cases}$

For the following $f(t)$, find $\mathscr{L}\{f(t)\}$ by evaluating the integral (2.1).

*3. $1 + (te^{-t} - 1)u(t)$.

4. $e^t u(t-1) + (1 - e^t)u(t-2)$.

5. $u(t) - 2u(t-1) + u(t-2)$.

6. $\sin t\{u(t) - u(t-2\pi)\}$.

2.3 Existence and uniqueness

The Laplace transform is an infinite-integral operator applied to the function $f(t)$. Not all functions can be integrated over even a finite range, and, as we have seen, not all infinite-range integrals converge. To ensure that the Laplace transform exists, $f(t)$ and s must be restricted in accordance with the following.

(i) $e^{-st}f(t)$ must be integrable for $0 \le t \le T < \infty$. It follows that $f(t)$ can have no nonintegrable singularity in $0 \le t \le T$, and a sufficient (but not necessary) condition is that $f(t)$ is *piecewise-continuous* in $0 \le t < \infty$.

(ii) The infinite integral must converge and do so *absolutely*. Thus

$$\int_0^\infty |e^{-st}f(t)|\,dt < \infty;$$

that is,

$$\int_0^\infty e^{-\sigma t}|f(t)|\,dt < \infty,$$

requiring that

$$\lim_{t\to\infty} e^{-\sigma t}|f(t)| = 0. \tag{2.10}$$

This is a necessary condition that limits the maximum rate of growth of $|f(t)|$ as $t \to \infty$.

If real constants C and a exist such that for all $t > T$

$$|f(t)| < Ce^{at},$$

$f(t)$ is said to be of *exponential order* as $t \to \infty$. Most functions of practical interest satisfy this condition. For example, t^n is of exponential order (any $a > 0$ will suffice), as is an exponential with a first-order polynomial in t as the exponent, but e^{t^2} is not. With this restriction, Eq. (2.10) is satisfied if σ is chosen appropriately.

Thus, for the Laplace transform to exist,

> $f(t)$ must be integrable in $0 \leq t \leq T < \infty$ (it is sufficient if it is *piecewise-continuous* there) and of *exponential order* as $t \to \infty$, with σ chosen to satisfy Eq. (2.10).

Under these conditions the transform integral converges uniformly (in s) as well as absolutely to a continuous function of s. Uniform convergence is a property of some infinite integrals and infinite series of functions (see, for example, Section 4.5), and in the case of the Laplace transform, allows differentiation with respect to s under the integral sign.

On the assumption that the preceding conditions are fulfilled, we will no longer cite σ when evaluating a transform.

In addition to the given conditions, there is another matter that is important when using the transform as a mathematical tool for the solution of a problem. Many problems (and certainly those appropriate to linear systems) are most easily solved in terms of the Laplace transform of the desired physical quantity $f(t)$, requiring the initial application of the transform and the subsequent recovery of $f(t)$ from a knowledge of $F(s)$. Are these two operations unique; that is, is there a one-to-one relationship between $f(t)$ and its transform $F(s)$?

With the restriction to piecewise-continuous functions $f(t)$, it is evident from the definition (2.1) of the Laplace transform that $F(s)$ is unique. In other words, to each $f(t)$ there corresponds one, and only one, transform $F(s)$. There is also an inverse transform, which we shall denote by the operator $\mathscr{L}^{-1}$, that maps a function of s into a function of t, thereby constituting the recovery operation. Is this also unique, or could several functions $f(t)$ have the same transform $F(s)$? Since we have not yet mathematically defined the inverse Laplace transform, and will not do so until later (see Chapter 7), it is difficult to derive sufficient conditions for a one-to-one relationship, but a necessary condition is evident from the following: If, in the process of Laplace transformation, information is destroyed, uniqueness is lost.

To see how information could be destroyed, we have only to examine the definition of the Laplace transform. Since the integration is from $t = 0$ to infinity, any information that $f(t)$ contains for $t < 0$ is suppressed by the transformation, and a one-to-one relationship between the s and t planes can exist only if $f(t) = 0$ for $t < 0$; that is,

$$f(t) \quad \text{is proportional to} \quad u(t).$$

Note that the presence of the unit-step function is not necessary for the determination of $\mathscr{L}\{f(t)\}$ but *is* necessary if we are to recover the original $f(t)$ from a knowledge of its transform.

2.4 Table of memorable transforms (TMT)

$$\underline{f(t) \qquad F(s)}$$

(i) $t^n u(t) \rightleftarrows n!/s^{n+1} \qquad (n = 0, 1, 2, \dots)$

(ii) $e^{-at}u(t) \rightleftarrows \dfrac{1}{s+a} \qquad (a = \text{any constant})$

(iii) $(\cos t)u(t) \rightleftarrows \dfrac{s}{s^2+1}$

(iv) $(\sin t)u(t) \rightleftarrows \dfrac{1}{s^2+1}$

These were all derived in the earlier examples and exercises, and the table can be read from right to left as well as from left to right. The reason we can get by with such a small table is that the Laplace transform has a number of general properties that can be used to extend the table indefinitely.

2.5 Properties of the Laplace transform

In the following, $F(s) = \mathscr{L}\{f(t)\}$, where $f(t)$ is assumed to satisfy the conditions given in Section 2.3. Since we are concerned only with the "direct" transform, it is not necessary to include $u(t)$ explicitly. The first four properties are particularly useful.

Linearity

$$\mathscr{L}\{c_1 f_1(t) + c_2 f_2(t)\} = c_1 F_1(s) + c_2 F_2(s), \tag{2.11}$$

where c_1 and c_2 are constants.

Proof: This follows immediately from the linearity of the integral operator, and the extension to any finite number of terms is obvious. □

Example. Determine $\mathscr{L}\{\sinh t\}$.

Since

$$\sinh t = \tfrac{1}{2}(e^t - e^{-t}),$$

we have

$$\mathscr{L}\{\sinh t\} = \frac{1}{2}\left(\frac{1}{s-1} - \frac{1}{s+1}\right) \qquad \text{(see TMT(ii) with } a = \pm 1\text{)}$$

$$= \frac{1}{s^2 - 1}.$$

Similarly,

$$\mathscr{L}\{\cosh t\} = \frac{s}{s^2 - 1}.$$

Scaling (or similarity)

$$\mathscr{L}\{f(at)\} = \frac{1}{a}F\left(\frac{s}{a}\right) \tag{2.12}$$

for real $a > 0$ (since the argument of f must be real and the function must vanish for negative arguments).

Proof:

$$\mathscr{L}\{f(at)\} = \int_0^\infty e^{-st}f(at)\,dt$$

$$= \frac{1}{a}\int_0^\infty e^{-(s/a)t'}f(t')\,dt' \qquad t' = at$$

$$= \frac{1}{a}F\left(\frac{s}{a}\right).$$

Note the multiplicative factor $1/a$, and observe how a appears in the denominator on one side of the equation and in the numerator on the other. □

Example. Determine $\mathscr{L}\{\sin 3t\}$.
From TMT(iv),

$$\mathscr{L}\{\sin 3t\} = \frac{1}{3}\frac{1}{(s/3)^2 + 1} = \frac{3}{s^2 + 9}.$$

s-Plane shift (or complex translation)

$$\mathscr{L}\{e^{at}f(t)\} = F(s - a) \tag{2.13}$$

for any a, real or complex.

Proof:

$$\mathscr{L}\{e^{at}f(t)\} = \int_0^\infty e^{-st}e^{at}f(t)\,dt$$
$$= \int_0^\infty e^{-(s-a)t}f(t)\,dt$$
$$= F(s-a). \quad \square$$

Example. Determine $\mathscr{L}\{e^{-t}t^2\}$.
Since

$$\mathscr{L}\{t^2\} = \frac{2}{s^3}, \qquad \text{TMT(i) with } n = 2$$

we have

$$\mathscr{L}\{e^{-t}t^2\} = \frac{2}{(s+1)^3}.$$

t-Plane shift (or real translation)

$$\mathscr{L}\{f(t-a)u(t-a)\} = e^{-as}F(s) \tag{2.14}$$

for real $a \geq 0$ (otherwise the shift displaces the function into $t < 0$).

Proof:

$$\mathscr{L}\{f(t-a)u(t-a)\} = \int_0^\infty e^{-st}f(t-a)u(t-a)\,dt$$
$$= \int_{+a}^\infty e^{-s(t'+a)}f(t')u(t')\,dt' \qquad t' = t - a$$
$$= \int_0^\infty e^{-s(t'+a)}f(t')\,dt'$$

since $u(t') = 0$ for $t' < 0$. Hence

$$\mathscr{L}\{f(t-a)u(t-a)\} = e^{-as}F(s). \quad \square$$

Example. Determine $\mathscr{L}\{(t-1)u(t-1)\}$.
Since

$$\mathscr{L}\{tu(t)\} = \frac{1}{s^2}, \qquad \text{TMT(i) with } n = 1$$

we have

$$\mathscr{L}\{(t-1)u(t-1)\} = \frac{e^{-s}}{s^2}. \tag{2.15}$$

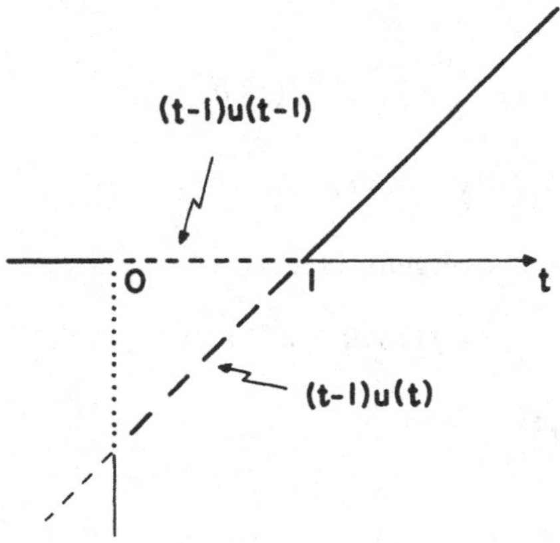

Fig. 2.8: The functions $(t-1)u(t-1)$ and $(t-1)u(t)$. The difference is confined to the range $0 \leq t \leq 1$, and is the source of the differences between the transforms in Eqs. (2.15) and (2.16).

The word "plane" in the description of this property is somewhat a misnomer, and is used to stress the analogy with the previous property, in which the word is meaningful. In either plane, multiplication by an exponential produces a shift in the other, but the t-plane shift formula requires special care in its use. The presence of a shift is indicated by the shifted argument of the step function, and in Eq. (2.14) the function being transformed is zero for $t < a$. Thus, $(t-1)u(t)$ is *not* a shifted function and, indeed, from TMT(i) with $n = 0$ and 1,

$$\mathscr{L}\{(t-1)u(t)\} = \frac{1}{s^2} - \frac{1}{s}. \tag{2.16}$$

On the other hand, $tu(t-1)$ *is* a shifted function, and since

$$\begin{aligned} tu(t-1) &= (t-1+1)u(t-1) \\ &= (t-1)u(t-1) + u(t-1), \end{aligned}$$

application of the formula gives

$$\mathscr{L}\{tu(t-1)\} = \left(\frac{1}{s^2} + \frac{1}{s}\right)e^{-s}.$$

Multiplication by t^n

$$\mathscr{L}\{t^n f(t)\} = (-1)^n \frac{d^n}{ds^n} F(s) \tag{2.17}$$

for $n = 0, 1, 2, \ldots$.

Proof:

$$(-1)^n \frac{d^n}{ds^n} F(s) = (-1)^n \frac{d^n}{ds^n} \int_0^\infty e^{-st} f(t)\,dt$$
$$= (-1)^n \int_0^\infty \frac{\partial^n}{\partial s^n}(e^{-st}) f(t)\,dt$$

since the integral is uniformly convergent. Hence

$$(-1)^n \frac{d^n}{ds^n} F(s) = \int_0^\infty e^{-st} t^n f(t)\,dt = \mathscr{L}\{t^n f(t)\}. \quad \square$$

Example. Determine $\mathscr{L}\{e^{-t}t^2\}$.

From TMT(ii) with $a = 1$,

$$\mathscr{L}\{e^{-t}t^2\} = (-1)^2 \frac{d^2}{ds^2}\left(\frac{1}{s+1}\right) = \frac{2}{(s+1)^3},$$

in agreement with the result previously obtained using the s-plane shift formula.

Division by t

$$\mathscr{L}\left\{\frac{f(t)}{t}\right\} = \int_s^\infty F(s')\,ds' \tag{2.18}$$

provided the transform on the left-hand side exists. A sufficient, but not necessary, condition is that $\lim_{t \to 0+}[f(t)]/t$ exist.

Proof:

$$\int_s^\infty F(s')\,ds' = \int_s^\infty \int_0^\infty e^{-s't} f(t)\,dt\,ds'$$
$$= \int_0^\infty \int_s^\infty e^{-s't} f(t)\,ds'\,dt$$
$$= \int_0^\infty e^{-st} \frac{f(t)}{t}\,dt = \mathscr{L}\left\{\frac{f(t)}{t}\right\}. \quad \square$$

Example. Determine $\mathscr{L}\{(1 - e^{-t})/t\}$.

From TMT(ii) with $a = 0$ and 1,

$$\mathscr{L}\{1 - e^{-t}\} = \frac{1}{s} - \frac{1}{s+1}$$

and hence

$$\mathscr{L}\left\{\frac{1 - e^{-t}}{t}\right\} = \ln \frac{s+1}{s}.$$

Periodicity

If $g(t)$ is the periodic function

$$g(t) = \sum_{n=0}^{\infty} f(t - n\tau)$$

of period τ, where

$$f(t) = f_1(t)\{u(t) - u(t - \tau)\},$$

then

$$\mathscr{L}\{g(t)\} = \frac{f(s)}{1 - e^{-\tau s}}. \tag{2.19}$$

Proof:

$$\mathscr{L}\{g(t)\} = \int_0^{\infty} e^{-st} \sum_{n=0}^{\infty} f(t - n\tau)\, dt$$

$$= \sum_{n=0}^{\infty} \int_0^{\infty} e^{-st} f_1(t - n\tau)\{u(t - n\tau) - u(t - \overline{n+1}\tau)\}\, dt$$

$$= \sum_{n=0}^{\infty} \int_0^{\infty} e^{-st} e^{-n\tau s} f_1(t)\{u(t) - u(t - \tau)\}\, dt$$

by a change of variable equivalent to the t-plane shift property. Hence

$$\mathscr{L}\{g(t)\} = \sum_{n=0}^{\infty} e^{-n\tau s} \int_0^{\infty} e^{-st} f(t)\, dt$$

$$= \frac{F(s)}{1 - e^{-\tau s}}$$

using the formula for an infinite geometric series. □

Example. Determine $\mathscr{L}\{g(t)\}$, where

$$g(t) = \sum_{n=0}^{\infty} f(t - 2n)$$

with $f(t) = u(t) - u(t - 1)$.

Since

$$F(s) = \int_0^1 e^{-st}\, dt = \frac{1 - e^{-s}}{s}$$

and the period τ of $g(t)$ is 2,

$$G(s) = \frac{1}{s}\frac{1 - e^{-s}}{1 - e^{-2s}} = \frac{1}{s(1 + e^{-s})}.$$

The given periodic function is a square wave.

Exercises

Using the TMT and the appropriate Laplace-transform properties, find $\mathscr{L}\{f(t)\}$ for the following $f(t)$.

*1. $e^t t^4$.

2. $(\sin t - \cos t)^2$.

3. $e^{-2t}\sin t$.

4. $e^{-t}(3\sinh 2t - 5\cosh 2t)$.

*5. $\sin 2(t-1)u(t-1)$.

6. $e^{2t}(t-1)^4 u(t-1)$.

7. $t^2 u(t-1)$.

8. $(\cos t)\{u(t) - u(t-2\pi)\}$.

9. $e^{-t}(\cos \pi t)u(t-1)$.

10. $\dfrac{\sinh t}{t}$.

11. $\dfrac{1}{t}(e^{-at} - e^{-bt})$.

12. $\dfrac{1}{t}(\cos at - \cos bt)$.

Find the Laplace transform of the periodic function $g(t) = \sum_{n=0}^{\infty} f(t - n\tau)$ having period τ and profile $f(t)$ as follows.

*13. $\tau = 2$, $f(t) = u(t) - 2u(t-1) + u(t-2)$ (square wave).

14. $\tau = 2$, $f(t) = t\{u(t) - u(t-1)\}$ (rectified sawtooth).

15. $\tau = 2$, $f(t) = t\{u(t) - u(t-2)\} - 2\{u(t-1) - u(t-2)\}$ (sawtooth).

16. $\tau = 2\pi$, $f(t) = (\sin t)\{u(t) - u(t-\pi)\}$ (rectified half sine wave).

2.6 Integration and differentiation

The importance of the Laplace transform for the solution of problems involving linear systems is due in large measure to these properties of the transform. To facilitate their presentation and subsequent application, it is convenient to introduce a simplified notation.

The *D* operator

For convenience and brevity we shall write

$$\frac{d}{dt} \quad \text{as } D.$$

Thus

$$\frac{d}{dt}f(t) = Df(t)$$

and since

$$\frac{d^2}{dt^2}f(t) = \frac{d}{dt}\frac{d}{dt}f(t),$$

this can be written as $D^2 f(t)$, and so on.

It is obvious that D is a linear-differential operator, as is $a_0 D + a_1 D$ where a_0 and a_1 are constants, but it is sufficient for our purposes to treat D only as a shorthand. In particular, we shall make no attempt to give a meaning to the operator independent of the function on which it acts, though we remark that this was the basis for the intuitive method that Heaviside developed.

Since integration is an operation that is the inverse of differentiation, this same symbol can be used to denote integration. If

$$Dg(t) = f(t)$$

then, formally at least,

$$g(t) = \frac{1}{D}f(t) = D^{-1}f(t),$$

and because

$$g(t) = \int^t f(t')\,dt',$$

it follows that

$$\int^t dt' \equiv D^{-1}.$$

Likewise, D^{-n} will denote an n-fold repeated integration. If the integral is definite with a specific lower limit, this will be shown as a subscript to the D^{-1} operator, so that

$$\int_a^t dt' = D_a^{-1}.$$

Integration

$$\mathscr{L}\{D_0^{-1}f(t)\} = \frac{F(s)}{s}. \tag{2.20}$$

Proof: Let $D_0^{-1}f(t) = g(t)$. Then

$$\mathscr{L}\{g(t)\} = \int_0^\infty e^{-st}g(t)\,dt$$

$$= \left[-\frac{e^{-st}}{s}g(t)\right]_0^\infty + \frac{1}{s}\int_0^\infty e^{-st}Dg(t)\,dt \qquad \text{by parts}$$

$$= \frac{F(s)}{s}$$

since $Dg(t) = f(t)$ and $g(0) = 0$. Symbolically,

$$\mathscr{L}\{D_0^{-1}f(t)\} = \frac{F(s)}{s},$$

and by repeated application of this result

$$\mathscr{L}\{D_0^{-n}f(t)\} = \frac{F(s)}{s^n} \tag{2.21}$$

for $n = 0, 1, 2, \ldots$. We observe that the integration generates positive powers of s in the denominator of the transform, that is, negative powers of s. □

Differentiation

If $Df(t)$ is piecewise-continuous for $t \geq 0$, implying that $f(t)$ is continuous,

$$\mathscr{L}\{Df(t)\} = sF(s) - f(0+), \tag{2.22}$$

where $f(0+) = \lim_{t\to 0+}f(t)$ is a right-hand limit.

Proof:

$$\mathscr{L}\{Df(t)\} = \int_0^\infty e^{-st}Df(t)\,dt$$

$$= \left[e^{-st}f(t)\right]_0^\infty + s\int_0^\infty e^{-st}f(t)\,dt \qquad \text{by parts}$$

$$= -f(0+) + sF(s).$$

By repeated application of this result, we can obtain the transforms of higher derivatives. Thus, if

$$g(t) = Df(t)$$

with $Dg(t)$ piecewise-continuous, then

$$g(0+) = f'(0+),$$

where the prime denotes the derivative, and

$$\begin{aligned}\mathscr{L}\{D^2f(t)\} &= \mathscr{L}\{Dg(t)\}\\ &= -g(0+) + s\mathscr{L}\{g(t)\}\\ &= -f'(0+) + s\{-f(0+) + sF(s)\}\\ &= -f'(0+) - sf(0+) + s^2F(s).\end{aligned}$$

In general, if $Df(t)$, $D^2f(t), \ldots, D^nf(t)$ are all piecewise-continuous,

$$\begin{aligned}\mathscr{L}\{D^nf(t)\} &= s^nF(s) - s^{n-1}f(0+) - s^{n-2}f^{(1)}(0+)\\ &\quad - \cdots - f^{(n-1)}(0+),\end{aligned} \tag{2.23}$$

where the limits are right hand and the superscripts in parentheses denote derivatives. It is important that the student find some way to remember this formula. More compactly,

$$\mathscr{L}\{D^nf(t)\} = s^nF(s) - \sum_{m=1}^{n} s^{n-m}f^{(m-1)}(0+),$$

and we observe that differentiation generates positive powers of s. □

2.7 Initial- and final-value theorems

These relate the initial ($t = 0+$) and final ($t = \infty$) values of $f(t)$ to the values of $sF(s)$ as $s \to \infty$ and $s \to 0$, respectively. As such, they can be used to obtain information about $f(t)$ from a knowledge of its transform and vice versa.

Initial-value theorem

If $Df(t)$ is piecewise-continuous for $t \geq 0$,

$$f(0+) = \lim_{s \to \infty} sF(s). \tag{2.24}$$

Proof: From Eq. (2.22)

$$sF(s) - f(0+) = \int_0^\infty e^{-st}Df(t)\,dt$$

and hence

$$\lim_{s \to \infty} sF(s) - f(0+) = \lim_{s \to \infty} \int_0^\infty e^{-st}Df(t)\,dt.$$

Since $Df(t)$ is piecewise-continuous for $0 \leq t < \infty$, by assumption the integral is uniformly convergent for Re s exceeding some finite number. We can therefore take the limit inside the integral sign on the right and the result is zero, giving

$$f(0+) = \lim_{s\to\infty} sF(s).$$

We remark that the theorem is also valid if $f(t)$ itself is piecewise-continuous (see Kaplan, 1962, pp. 336–8). □

Final-value theorem

If $Df(t)$ is piecewise-continuous for $t \geq 0$ and $\lim_{t\to\infty} f(t)$ exists,

$$\lim_{t\to\infty} f(t) = \lim_{s\to 0} sF(s). \tag{2.25}$$

Proof: We again start with Eq. (2.22), but this time take the limit as $s \to 0$. Then

$$\begin{aligned}\lim_{s\to 0} sF(s) - f(0+) &= \lim_{s\to 0} \int_0^\infty e^{-st} Df(t)\,dt \\ &= \int_0^\infty Df(t)\,dt = [f(t)]_0^\infty \\ &= \lim_{t\to\infty} f(t) - f(0+),\end{aligned}$$

giving

$$\lim_{t\to\infty} f(t) = \lim_{s\to 0} sF(s).$$

This theorem is also valid if $f(t)$ is merely piecewise-continuous.

The restriction that $\lim_{t\to\infty} f(t)$ must exist (i.e., is finite and unique) eliminates functions such as e^t, which tends to infinity as $t \to \infty$, and $\cos t$, whose limit is undefined; since the main use of the theorem is to obtain the limiting value of $f(t)$ from a knowledge of $F(s)$, it is not too helpful to express the restriction in this way. □

As is evident from the preceding proof, we can get at the restriction differently by considering $F(s)$. In general, a Laplace-transform integral converges only in a half-plane $\sigma > a$, and the point $s = 0$ may be neither in this region nor on the boundary. In order for the required limit in the s plane to exist, we must be able to reach $s = 0$ from the half-plane of convergence of the Laplace transform, implying that the only singularities of $sF(s)$ are in the left-half s plane, where $\sigma < 0$.

Exercises

For the following $f(t)$ determine $F(s)$ and then verify the initial-value theorem.

1. $3 + \cos 2t$.
2. $2\cos t + te^{-t}\sin t$.
3. $e^{2t}u(t-1)$.

For the following $f(t)$ determine $F(s)$ and then verify the final-value theorem.

4. t^2e^{-2t}.
5. $1 + e^{-t}(\cos t + \sin t)$.
6. $tu(t) - (t-2)u(t-1)$.

2.8 Application to differential equations

The Laplace transform is an effective tool for solving initial-value problems associated with ordinary linear-differential equations with constant coefficients, where the initial conditions are prescribed at $t = 0$. In contrast to the method of exponential substitution, in which it is necessary to obtain the general solution first and then differentiate to impose the initial conditions, the Laplace-transform method imposes these conditions at the outset.

Consider the nth-order differential equation

$$\left(a_0D^n + a_1D^{n-1} + \cdots + a_n\right)x(t) = f(t), \tag{2.26}$$

where $a_0, a_1, \ldots, a_n$ are constants with $a_0 \neq 0$. Given $f(t)$ the task is to determine $x(t)$ subject to prescribed values of $x(0+), x'(0+), \ldots, x^{(n-1)}(0+)$, where the prime(s) and bracketed superscripts denote derivatives. These are sufficient to ensure a unique solution, and with the understanding that the prescribed values are all right-hand limits, we shall henceforth omit the postscripts $+$. If the constraints are specified at some time other than zero (e.g., $t = t_1$) it is necessary to shift the origin of time to $t = t_1$ before applying a Laplace transform.

To ensure a one-to-one relationship between each function of t and its transform, it is assumed that all functions are zero for $t < 0$.

Example 1. Solve $(D^2 + 4D + 3)x(t) = 0$ with $x(0) = 2$ and $x'(0) = -4$.

The task is to find $x(t)$, which is assumed zero for $t < 0$. If $X(s) = \mathscr{L}\{x(t)\}$, as always, application of a Laplace transform to the left-hand side of the differential equation gives

$$\begin{aligned}\mathscr{L}\{(D^2 + 4D + 3)x(t)\} &= \mathscr{L}\{D^2x(t)\} + 4\mathscr{L}\{Dx(t)\}\\ &\quad + 3\mathscr{L}\{x(t)\}\\ &= s^2X(s) - sx(0) - x'(0)\\ &\quad + 4(sX(s) - x(0)) + 3X(s)\end{aligned}$$

using the derivative property of Eq. (2.22). On inserting the initial values,

$$\begin{aligned}\mathscr{L}\{(D^2 + 4D + 3)x(t)\} &= (s^2 + 4s + 3)X(s) - 2s - (-4) + 4(-2)\\ &= (s+1)(s+3)X(s) - 2s - 4,\end{aligned}$$

and since the Laplace transform of the right-hand side (zero) is zero,

$$(s+1)(s+3)X(s) - 2s - 4 = 0,$$

giving

$$X(s) = \frac{2s+4}{(s+1)(s+3)}. \tag{2.27}$$

This is the Laplace transform of the desired solution $x(t)$ and is a rational function of s, in which the numerator is contributed by the initial values and the denominator is obtained from the differential operator.

We note in passing that the initial-value theorem applied to Eq. (2.27) confirms that $x(0) = 2$, and the final-value theorem shows that $\lim_{t\to\infty} x(t) = 0$.

In the context of our treatment so far, the inverse operation to recover $x(t)$ can be accomplished only if $x(t)$ can be recognized using the TMT and the Laplace-transform properties of Sections 2.5 and 2.6. Although it is not immediately evident how this can be done, an alert reader might observe that

$$X(s) = \frac{s+1+s+3}{(s+1)(s+3)} = \frac{1}{s+3} + \frac{1}{s+1}.$$

Hence, from TMT(ii) with $a = 3$ and $a = 1$,

$$x(t) = (e^{-3t} + e^{-t})u(t)$$

and, in accordance with our assumption, $x(t) = 0$ for $t < 0$.

Example 2. Solve $(D^2 + 1)x(t) = f(t)$ with $f(t) = 2\cos t$ and $x(0) = 1$, $x'(0) = 2$.

Assuming $x(t)$ and $f(t)$ are zero for $t < 0$, application of a Laplace transform to the differential equation gives

$$\begin{aligned}\mathscr{L}\{(D^2+1)x(t)\} &= s^2X(s) - sx(0) - x'(0) + X(s)\\ &= (s^2+1)X(s) - s - 2\\ &= \mathscr{L}\{2\cos t\}\\ &= \frac{2s}{s^2+1}\end{aligned}$$

from TMT(iii). Hence

$$(s^2+1)X(s) = s + 2 + \frac{2s}{s^2+1}$$

giving

$$X(s) = \frac{s+2}{s^2+1} + \frac{2s}{(s^2+1)^2} = \frac{(s^2+1)(s+2)+2s}{(s^2+1)^2},$$

which is again a rational function of s. Even an alert reader may now be close to despair, but since

$$\mathscr{L}^{-1}\left\{\frac{s}{s^2+1}\right\} = (\cos t)u(t), \qquad \text{TMT(iii)}$$

and

$$\mathscr{L}^{-1}\left\{\frac{1}{s^2+1}\right\} = (\sin t)u(t), \qquad \text{TMT(iv)}$$

with

$$\frac{2s}{(s^2+1)^2} = -\frac{d}{ds}\frac{1}{s^2+1},$$

use of the multiplication by t^n property with $n = 1$ gives

$$x(t) = (\cos t + 2\sin t + t\sin t)u(t).$$

It can be verified that this satisfies all the conditions of the problem and, in accordance with our assumption, $x(t) = 0$ for $t < 0$.

We were able to complete this solution because of a fortunate grouping of the terms in the rational-function expression for $X(s)$. It is too much to hope that this will always happen, and what is needed is a systematic procedure for splitting up a rational function. This is discussed next.

2.9 Partial fractions

A rational function $X(s)$ is the ratio of two polynomials in s. Without loss of generality we can take the coefficient of the highest power of s in the denominator to be unity, and thus

$$X(s) = \frac{P(s)}{Q(s)} \tag{2.28}$$

with

$$P(s) = b_0 s^m + b_1 s^{m-1} + \cdots + b_m,$$

$$Q(s) = s^n + a_1 s^{n-1} + \cdots + a_n.$$

If $b_0 \neq 0$, the degrees of the polynomials $P(s)$ and $Q(s)$ are m and n, respectively, and by analogy with fractions, the function is called *proper* if $m < n$ and *improper* if $m \geq n$. Since, for an improper rational function, we can always divide $Q(s)$ into $P(s)$ to produce a proper rational function plus a polynomial; for example,

$$\frac{s^5 - s^4 + 1}{s^4 + s^3} = s - 2 + \frac{2s^3 + 1}{s^4 + s^3},$$

it is sufficient to consider only proper rational functions, for which $m < n$.

Any polynomial of degree n has n zeros, possibly complex, and if the zeros of $Q(s)$ are $s_1, s_2, \ldots, s_n$, the rational function $X(s)$ can be written as

$$X(s) = \frac{P(s)}{(s - s_1)(s - s_2) \cdots (s - s_n)}.$$

A partial-fraction expansion is an expression of this as a sum of elementary proper rational functions called *partial fractions* and is possible only if $X(s)$ is proper.

Determination

If the zeros $s_1, s_2, \ldots, s_n$ of the denominator polynomial in Eq. (2.28) are all distinct, that is, no two are the same, the partial-fraction representation of $X(s)$ is

$$X(s) = \frac{A_1}{s - s_1} + \frac{A_2}{s - s_2} + \cdots + \frac{A_n}{s - s_n}, \tag{2.29}$$

where $A_1, A_2, \ldots, A_n$ are constants. In any given case, one or more constants may be zero; for example, $A_1 = 0$ if s_1 is also a zero of the numerator polynomial $P(s)$. If, on the other hand, any one of the zeros $s_1, s_2, \ldots, s_n$ is repeated, there is a corresponding number of terms in the partial-fraction representation. Thus if the zero s_1 occurs k times (i.e., is a kth-order zero), but all other zeros are distinct,

$$X(s) = \frac{A_1}{s - s_1} + \frac{A_2}{(s - s_1)^2} + \cdots + \frac{A_k}{(s - s_1)^k} + \frac{A_{k+1}}{s - s_{k+1}}$$

$$+ \cdots + \frac{A_n}{s - s_n}. \tag{2.30}$$

In both cases the number of terms is equal to the degree of the denominator polynomial $Q(s)$.

These representations can be validated by putting the right-hand sides of Eqs. (2.29) and (2.30) over a common denominator and comparing the resulting numerators with $P(s)$. The process of "multiplying up" and equating the coefficients of like powers of s can also be used to determine the coefficients, but this is not recommended. It is too tedious and conducive to error, and there is a far more efficient procedure available.

We consider first the case when all the zeros $s_1, s_2, \ldots, s_n$ are first order (simple) and therefore distinct. To determine the coefficient A_1 in Eq. (2.29), multiply through by $s - s_1$ to give

$$(s - s_1)X(s) = A_1 + (s - s_1)\left\{\frac{A_2}{s - s_2} + \cdots + \frac{A_n}{s - s_n}\right\}.$$

On taking the limit as $s \to s_1$, all terms on the right-hand side vanish except A_1, and thus

$$A_1 = \lim_{s \to s_1} \{(s - s_1)X(s)\}.$$

The other coefficients can be found in a similar manner, and in general

$$A_r = \lim_{s \to s_r} \{(s - s_r)X(s)\} \tag{2.31}$$

for $r = 1, 2, \ldots, n$. Since $X(s) = [P(s)]/[Q(s)]$, an equivalent statement of this result is

$$A_r = \frac{P(s_r)}{Q'(s_r)}, \tag{2.32}$$

where the prime denotes the derivative.

Example 1. Express in partial-fraction form $X(s) = (s^2 + 5)/(s^3 + 2s^2 - s - 2)$.

The rational function is proper and since

$$X(s) = \frac{s^2 + 5}{(s-1)(s+1)(s+2)},$$

all zeros of the denominator are distinct. The required representation is then

$$X(s) = \frac{A_1}{s-1} + \frac{A_2}{s+1} + \frac{A_3}{s+2},$$

where

$$A_1 = \lim_{s \to 1} \left\{ \frac{s^2 + 5}{(s+1)(s+2)} \right\} = 1,$$

$$A_2 = \lim_{s \to -1} \left\{ \frac{s^2 + 5}{(s-1)(s+2)} \right\} = -3,$$

and

$$A_3 = \lim_{s \to -2} \left\{ \frac{s^2 + 5}{(s-1)(s+1)} \right\} = 3.$$

Hence

$$X(s) = \frac{1}{s-1} - \frac{3}{s+1} + \frac{3}{s+2}.$$

The calculation is easily accomplished on a single line by using the "finger" method. We observe that A_1 is obtained by covering (with a finger) the corresponding factor in the denominator of $X(s)$ and putting $s = 1$ in the portion still visible. Similarly, A_2 is obtained by covering the factor $s + 1$, and A_3 by covering $s + 2$. Thus

$$X(s) = \frac{6/(2 \cdot 3)}{s-1} + \frac{6/(-2 \cdot 1)}{s+1} + \frac{9/(-3 \cdot -1)}{s+2}$$
$$= \frac{1}{s-1} - \frac{3}{s+1} + \frac{3}{s+2}.$$

We now consider the case when one (or more) of the zeros $s_1, s_2, \ldots, s_n$ is repeated. If, for example, s_1 is a kth-order zero and all others are distinct, the representation of $X(s)$ is as shown in Eq. (2.30), which we rewrite as

$$X(s) = \frac{B_0}{(s-s_1)^k} + \frac{B_1}{(s-s_1)^{k-1}} + \cdots + \frac{B_{k-1}}{s-s_1} + \sum_{p=k+1}^{n} \frac{A_p}{s-s_p}. \tag{2.33}$$

Since the As are all associated with first-order zeros, they can be obtained by the above method, and the task that remains is to find the Bs.

Multiplying through by $(s - s_1)^k$, we have

$$(s - s_1)^k X(s) = B_0 + B_1(s - s_1) + \cdots + B_{k-1}(s - s_1)^{k-1}$$
$$+ (s - s_1)^k \sum_{p=k+1}^{n} \frac{A_p}{s - s_p}, \tag{2.34}$$

from which B_0 follows immediately:

$$B_0 = \lim_{s \to s_1} \left\{ (s - s_1)^k X(s) \right\}.$$

To find B_1, differentiate Eq. (2.34) with respect to s, giving

$$\frac{d}{ds}\left[(s - s_1)^k X(s)\right] = B_1 + 2B_2(s - s_1)$$
$$+ \cdots + (k - 1) B_{k-1}(s - s_1)^{k-2}$$
$$+ \frac{d}{ds}\left[(s - s_1)^k \sum_{p=k+1}^{n} \frac{A_p}{s - s_p}\right],$$

from which we have

$$B_1 = \lim_{s \to s_1} \frac{d}{ds}\left[(s - s_1)^k X(s)\right],$$

and in general

$$B_r = \lim_{s \to s_1} \left\{ \frac{1}{r!} \frac{d^r}{ds^r}\left[(s - s_1)^k X(s)\right] \right\} \tag{2.35}$$

for $r = 0, 1, 2, \ldots, k - 1$.

Example 2. Express in partial-fraction form $X(s) = (3s - 1)/[s(s - 1)^3]$.

Since $s = 1$ is a third-order zero, the required representation is

$$X(s) = \frac{B_0}{(s - 1)^3} + \frac{B_1}{(s - 1)^2} + \frac{B_2}{s - 1} + \frac{1}{s},$$

where we have already used the finger method to find the coefficient of $1/s$. From Eq. (2.35)

$$B_0 = \lim_{s \to 1} \left\{ \frac{3s - 1}{s} \right\} = 2,$$

$$B_1 = \lim_{s \to 1} \frac{1}{1!} \frac{d}{ds} \left\{ \frac{3s - 1}{s} \right\} = \lim_{s \to 1} \left\{ \frac{1}{s^2} \right\} = 1,$$

$$B_2 = \lim_{s \to 1} \frac{1}{2!} \frac{d^2}{ds^2} \left\{ \frac{3s - 1}{s} \right\} = \lim_{s \to 1} \left\{ -\frac{1}{s^3} \right\} = -1,$$

and hence

$$X(s) = \frac{2}{(s-1)^3} + \frac{1}{(s-1)^2} - \frac{1}{s-1} + \frac{1}{s}.$$

The finger method is also applicable when a zero is repeated. In the above example we observe that the Bs are all obtained by covering the factor $(s-1)^3$ in the denominator of $X(s)$. For the highest power the coefficient (B_0) is found by simply putting $s = 1$ in the portion still visible; for the next-highest power we have to differentiate what is visible, divide by $1!$ $(= 1)$, and then put $s = 1$; for the next, differentiate twice, divide by $2!$, and put $s = 1$; and so on. A systematic procedure such as this eliminates the need to memorize Eq. (2.35).

Exercises

Express the following $X(s)$ in partial-fraction form.

*1. $\dfrac{s}{(s+1)(s+2)}$.

2. $\dfrac{3s^2 - s + 6}{s(s+1)(s+2)(s+3)}$.

3. $\dfrac{4s^2 - 1}{(s^2-1)(s^2-4)}$.

4. $\dfrac{3s(s+3)}{(2s+1)(s+1)(s+2)}$.

5. $\dfrac{1}{s^2(s^2-1)}$.

6. $\dfrac{6s^2+1}{s^3(s+1)}$.

*7. $\dfrac{4}{s^2(s+1)(s+2)}$.

8. $\dfrac{32}{(s^2-1)^4}$.

9. $\dfrac{4s+3}{s^3(s+1)}$.

10. $\dfrac{4}{s^4-1}$.

Application

The Laplace-transform solutions $X(s)$ of the differential equations considered in Section 2.8 are all proper rational functions of s, and expansion in partial fractions can then assist in obtaining $x(t)$, that is, in deriving the inverse transform $\mathscr{L}^{-1}\{X(s)\}$. The expansion guarantees that the inverse can be found from TMT(i) or (ii) with, at most, the assistance of the s-plane shift property. If all of the zeros $s_1, s_2, \ldots, s_n$ of the denominator polynomial are distinct, then, from Eq. (2.32),

$$X(s) = \sum_{r=1}^{n} \frac{P(s_r)}{Q'(s_r)} \frac{1}{s - s_r} \qquad (2.36)$$

and, from TMT(ii),

$$x(t) = \sum_{r=1}^{n} \frac{P(s_r)}{Q'(s_r)} e^{s_r t} u(t), \tag{2.37}$$

a result known as Heaviside's expansion formula. If a zero is repeated, any term such as $(s - s_r)^{-m}$ in the partial-fraction expansion can be inverted using TMT(i) and the s-plane shift property to give

$$\frac{1}{(m-1)!} t^{m-1} e^{s_r t} u(t).$$

On the other hand, expansion should be used judiciously and only to the extent necessary to express $X(s)$ in a form whose inverse can be recognized. In particular, the inverse of a real function of s is real, and if $X(s)$ is real, then $x(t)$ should be presented in a real form.

A few examples may serve to illustrate these points. Consider

$$X(s) = \frac{1}{s^2+1}.$$

Most readers will recognize this as a memorable transform and, from TMT(iv),

$$x(t) = (\sin t)u(t).$$

Alternatively, because

$$(s^2+1) = (s-i)(s+i),$$

partial-fraction expansion can always be employed to give

$$X(s) = \frac{1}{2i}\left(\frac{1}{s-i} - \frac{1}{s+i}\right),$$

which then requires that the exponential functions of t be combined as the sine function. A somewhat similar example is

$$X(s) = \frac{1}{(s^2+1)(s^2+4)}.$$

This could be expressed as a sum of four partial fractions, but since the zeros of the denominator are purely imaginary, it is better to avoid this. Recognizing that the entire expression is a function of s^2, we have

$$X(s) = \frac{1}{(y+1)(y+4)},$$

which is a rational function of $y = s^2$. Hence

$$X(s) = \frac{1}{3}\left(\frac{1}{y+1} - \frac{1}{y+4}\right) = \frac{1}{3}\left(\frac{1}{s^2+1} - \frac{1}{s^2+4}\right)$$

whose inverse is known. Similarly, if

$$X(s) = \frac{s}{(s^2+1)(s^2+4)},$$

the preceding procedure applied to $[X(s)]/s$ gives

$$X(s) = \frac{1}{3}\left(\frac{s}{s^2+1} - \frac{s}{s^2+4}\right),$$

which can be inverted using TMT(iii) and the scaling property.

Finally, a note of caution:

$$X(s) = \frac{e^{-s}}{s(s+1)}$$

is *not* a rational function, but since the part without the exponential *is*, the standard process can be applied to this part, giving

$$X(s) = e^{-s}\left(\frac{1}{s} - \frac{1}{s+1}\right).$$

From TMT(ii) and the t-plane shift property,

$$x(t) = (1 - e^{-(t-1)})u(t-1).$$

Exercises

By expanding in partial fractions (or otherwise), use the TMT to find $x(t)$ for the following $X(s)$.

*1. $\dfrac{2}{s^2-1}$.

2. $\dfrac{6s}{(s-1)(s^2-4)}$.

3. $\dfrac{1}{s^2(s^2+1)}$.

4. $\dfrac{3(s^2+2)}{(s^2+1)(s^2+4)}$.

*5. $\dfrac{1}{s(s^2+1)}$. [*Hint*: Multiply and divide by s.]

6. $\dfrac{72}{s(s^2+1)(s^2+9)}$.

7. $\dfrac{s-2}{s^4(s-1)}$.

8. $\dfrac{4(s^3+1)}{(s^2-1)^2}$.

9. $\dfrac{s+1}{s^2+2s+5}$.

10. $\dfrac{2}{(s-1)(s^2+1)}$. [*Hint*: Multiply and divide by $s+1$.]

*11. $\dfrac{e^{-2s}}{(s-1)(s-2)}$.

12. $\dfrac{2s-e^{-s}}{(s+1)^2}$.

2.10 Laplace-transform solution of differential equations

We now return to the solution of an nth-order differential equation of the form

$$(a_0D^n + a_1D^{n-1} + \cdots + a_n)x(t) = f(t), \tag{2.38}$$

where $a_0, a_1, \ldots, a_n$ are constants with $a_0 \neq 0$. The function $f(t)$ is assumed piecewise-continuous and we distinguish two cases:

(i) $f(t) = f_1(t)$, where $f_1(t)$ is piecewise-continuous in $0 < t < \infty$ and continuous with all derivatives continuous in $-\infty < t \leq 0$, and

(ii) $f(t) = f_1(t)u(t)$.

In the latter case $f(t) = 0$ for $t < 0$ and has a jump discontinuity at $t = 0$ if $f_1(0) \neq 0$. We seek the solution $x(t)$ of Eq. (2.38) subject to the prescribed values of $x, x', \ldots, x^{(n-1)}$ at $t = 0$.

The Laplace-transform method of solution is based on the assumption that $x(t)$ and $f(t)$ are zero for $t < 0$, and produces a solution consistent with this assumption. As evident from the second example in Section 2.8, the application of a Laplace transform to the differential equation gives

$$V(s)X(s) - A(s) = F(s), \tag{2.39}$$

where, as always, $X(s)$ and $F(s)$ are the Laplace transforms of $x(t)$ and $f(t)$. The function $A(s)$ is a polynomial of degree $\leq n - 1$ produced by the prescribed values at $t = 0$, and is zero if these values are zero. Also

$$V(s) = a_0s^n + a_1s^{n-1} + \cdots + a_n, \tag{2.40}$$

which is simply the *characteristic function* whose zeros are the characteristic roots of the differential equation.

By rearranging the terms in Eq. (2.39),

$$X(s) = Y(s) + Z(s), \tag{2.41}$$

where

$$Y(s) = \frac{A(s)}{V(s)}$$

is a proper rational function, and

$$Z(s) = \frac{F(s)}{V(s)}$$

is typically a combination of proper rational functions and exponentials associated with any step functions located at times $t > 0$. If $y(t)u(t) = \mathscr{L}^{-1}\{Y(s)\}$ and $z(t)u(t) = \mathscr{L}^{-1}\{Z(s)\}$, the solution obtained by the

Laplace transform method is then

$$x(t) = y(t)u(t) + z(t)u(t). \tag{2.42}$$

This is, of course, zero for $t < 0$ in accordance with the requirement for a one-to-one relationship between a function and its transform.

If the problem is viewed as an initial-value problem, the solution is of interest only at times t subsequent to that when the initial values are imposed, and since $u(t) = 1$ for $t > 0$,

$$x(t) = y(t) + z(t), \qquad t > 0.$$

However, there are also applications in which the solution is required for $t < 0$ as well, and regardless of the application, it is mathematically legitimate to seek the solution of Eq. (2.38) for all t, $-\infty < t < \infty$. The correct solution is not necessarily zero for $t < 0$, and is certainly not if $y(t) \neq 0$, but the solution for $t < 0$ (and, hence, for all t) can be *deduced* from that in Eq. (2.42).

The Laplace-transform solution $x(t)$ has two distinct parts. The first is $y(t)u(t)$. This incorporates the effect of the prescribed (initial) values and for $t > 0$ is a solution of the homogeneous equation satisfying the conditions at $t = 0+$. But any solution of the equation must be continuous, along with its first $n - 1$ derivatives, and is uniquely defined by their values at any t. Thus, the prescribed conditions at $t = 0$ are likewise those defining the solution for $t < 0$, and the solution satisfying these is $y(t)$. It is therefore the solution for all t.

The second part of Eq. (2.42) is $z(t)u(t)$, for which the initial values are obviously zero; that is,

$$z(0) = z'(0) = \cdots = z^{(n-1)}(0) = 0,$$

implying that $z(t)u(t)$ and its first $n - 1$ derivatives are continuous at $t = 0$. This is so in spite of the (possible) jump discontinuity in $f(t)$ at $t = 0$, and if $f(t) = 0$ for $t < 0$ (case (ii)), $z(t)u(t)$ is the required solution for all t. But if $f(t) = f_1(t) \neq 0$ for $t < 0$ (case (i)), the solution must be nonzero for $t < 0$, and the same process of continuation as that used above defines the solution as $z(t)$. It is therefore the solution for all t.

Thus, the solution for all t, $-\infty < t < \infty$, is

$$x(t) = y(t) + \begin{cases} z(t) & \text{if } f(t) = f_1(t), \\ z(t)u(t) & \text{if } f(t) = f_1(t)u(t). \end{cases} \tag{2.43}$$

We emphasize that this is a *deduction* from the Laplace-transform solution of Eq. (2.42). It is based on the mathematical properties of a solution of Eq. (2.38), and since the (one-sided) Laplace transform cannot take into account values other than zero for $t < 0$, the process must be used with care. Thus, if $f(t) = 1 + e^{-t}u(t+1)$, the Laplace-transform solution is obtained for $f(t) = (1 + e^{-t})u(t)$, and the solution deduced for all t does not include the effect of the jump at $t = -1$. In order to treat the given input, it is necessary to shift the origin of time to put the discontinuity in $f(t)$ into the region $t > 0$, using arbitrary initial values that are then chosen to satisfy the original conditions. Fortunately, such cases are of no concern to us.

Example. Use the Laplace-transform technique to find $x(t)$ where

$$(D^2 + 3D + 2)x(t) = f(t)$$

with $f(t) = 2\{u(t) - u(t-1)\}$ and $x(0) = 1$, $x'(0) = 2$. Deduce the solution for all t.

Assuming $x(t)$ is zero for $t < 0$, the application of a Laplace transform gives

$$\begin{aligned}\mathscr{L}\{(D^2 + 3D + 2)x(t)\} &= s^2X(s) - sx(0) - x'(0) \\ &\quad + 3\{sX(s) - x(0)\} + 2X(s) \\ &= (s^2 + 3s + 2)X(s) - s - 2 - 3 \\ &= (s+1)(s+2)X(s) - s - 5 \\ &= 2\mathscr{L}\{u(t) - u(t-1)\} \\ &= \frac{2}{s}(1 - e^{-s}).\end{aligned}$$

Hence

$$\begin{aligned}X(s) &= \frac{s+5}{(s+1)(s+2)} + \frac{2}{s(s+1)(s+2)}(1 - e^{-s}) \\ &= \frac{4}{s+1} - \frac{3}{s+2} + \left(\frac{1}{s} - \frac{2}{s+1} + \frac{1}{s+2}\right)(1 - e^{-s}),\end{aligned}$$

giving

$$\begin{aligned}x(t) &= (4e^{-t} - 3e^{-2t})u(t) + (1 - 2e^{-t} + e^{-2t})u(t) \\ &\quad - (1 - 2e^{-(t-1)} + e^{-2(t-1)})u(t-1).\end{aligned}$$

We recognize the first term on the right-hand side as $y(t)u(t)$ and the last two terms as $z(t)u(t)$. The solution for all t is then

$$\begin{aligned}x(t) &= 4e^{-t} - 3e^{-2t} + (1 - 2e^{-t} + e^{-2t})u(t) \\ &\quad - (1 - 2e^{-(t-1)} + e^{-2(t-1)})u(t-1).\end{aligned}$$

On the other hand, if the given function $f(t)$ had been

$$f(t) = 2\{1 - u(t-1)\},$$

the solution for all t would have been

$$\begin{aligned} x(t) &= 4e^{-t} - 3e^{-2t} + 1 - 2e^{-t} + e^{-2t} \\ &\quad -(1 - 2e^{-(t-1)} + e^{-2(t-1)})u(t-1) \\ &= 1 + 2e^{-t} - 2e^{-2t} - (1 - 2e^{-(t-1)} + e^{-2(t-1)})u(t-1). \end{aligned}$$

Exercises

For the following differential equations use the Laplace-transform technique to find $x(t)$ and then (first 10 problems only) deduce the solution for all t, $-\infty < t < \infty$.

*1. $(D^2 + 4)x(t) = 3\cos t$, with $x(0) = 0$, $x'(0) = 2$.

2. $(D^3 + D^2 - 4D - 4)x(t) = 12e^t$, with $x(0) = -2$, $x'(0) = 2$, $x''(0) = -2$.

3. $(D^3 - 2D^2 - D + 2)x(t) = 6u(t)$, with $x(0) = x'(0) = x''(0) = 0$.

4. $(D^2 + 3D + 2)x(t) = 4tu(t)$, with $x(0) = 0$, $x'(0) = 1$.

*5. $(D + 1)x(t) = 2(\sin t)u(t)$, with $x(0) = 2$.

6. $(D^2 + 2D + 5)x(t) = 8te^{-t}$, with $x(0) = 1$, $x'(0) = 3$.

7. $(D^2 + 2D + 1)x(t) = te^{-t}u(t)$, with $x(0) = 2$, $x'(0) = 1$.

8. $(D^3 + 3D^2 + 2D)x(t) = 6e^t$, with $x(0) = 3$, $x'(0) = 0$, $x''(0) = 2$.

9. $(D^3 + D)x(t) = 6(\sin 2t)u(t)$, with $x(0) = 1$, $x'(0) = 0$, $x''(0) = 2$.

10. $(D^3 + D^2)x(t) = t - (t-2)u(t-2)$, with $x(0) = 0$, $x'(0) = 2$, $x''(0) = 0$.

*11. $(D^2 + 4D + 3)x(t) = 4\sin t + 8\cos t$, with $x(0) = 1$, $x'(0) = 3$.

12. $(D^2 + 2D + 2)x(t) = 3e^{-t}(\sin 2t)u(t)$, with $x(0) = 2$, $x'(0) = -1$.

13. $(D^3 + 1)x(t) = -6e^{-t}u(t)$, with $x(0) = 0$, $x'(0) = x''(0) = 1$.

14. $(D + 4)x(t) + 3D_0^{-1}x(t) = 2e^{-t}u(t)$, with $x(0) = -1$.

15. $(D - 1)^2x(t) = t\{u(t) - u(t-1)\}$, with $x(0) = 0$, $x'(0) = 3$.

CHAPTER 3

Linear systems

Having now arrived at the heart of our subject, we show how the mathematical techniques that have been discussed can be used to solve some of the problems associated with linear systems.

3.1 Basic concepts

In its most general form a system is any physical device that when stimulated or excited, produces a response. The system could be as complicated as (a model of) the human brain or as simple as an electrical circuit with a lumped resistance, and the problem the engineer could face is to determine the response or output of the system resulting from a known excitation or input. To develop procedures that are applicable to a variety of systems, it is necessary to restrict the type of system considered.

Physical description

For the systems of concern to us it is assumed that the input and output are functions of a single real variable t, which we shall speak of as time, and that a causal relationship exists between the two. The term *causal* implies that the output is a function of the input alone, and because of this, the system is "nonanticipatory"; that is, there is no output until the input is applied. This is an attribute of all physically realizable systems. For convenience we shall refer to the input and output as signals, and if these are denoted by $f(t)$ and $x(t)$, respectively, a system can be depicted as shown in Fig. 3.1.

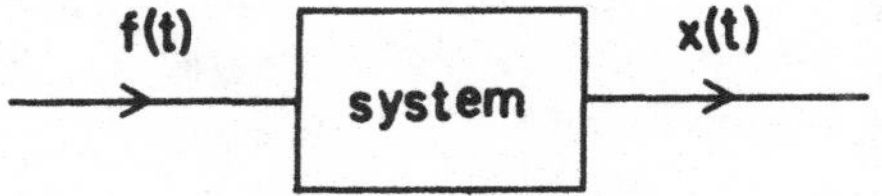

Fig. 3.1: A system.

There are two restrictions that will be imposed. The first is *linearity*. The necessary and sufficient condition for a system to be linear is that if an input $f_1(t)$ produces the output $x_1(t)$, that is,

$$f_1(t) \to x_1(t),$$

and similarly

$$f_2(t) \to x_2(t),$$

then

$$c_1 f_1(t) + c_2 f_2(t) \to c_1 x_1(t) + c_2 x_2(t)$$

for any constants c_1 and c_2. This is the principle of superposition, and the extension to any finite number of inputs follows immediately. Although few physical systems are linear under all circumstances (even a circuit could cease to be linear if the applied voltage were large enough to overheat a resistor), many are linear to an adequate degree subject, for example, to some reasonable limitation on the magnitude of the input.

The second restriction is time-invariance. A system is said to be *time-invariant* if a time-shifted input yields the corresponding time-shifted output, that is, if

$$f(t) \to x(t)$$

implies

$$f(t+\tau) \to x(t+\tau)$$

for all τ. As the name indicates, a time-invariant system is one whose properties do not change with time, and no system could satisfy this completely. After all, any system must have been created at some time in the past and will cease to exist at some time in the future. Even if we confine our attention to a more realistic time scale, systems do change with time as their components age, so that time-invariance is an ideal that is seldom achieved. Nevertheless, over restricted periods of time many systems are time-invariant to a reasonable extent, and this fact, coupled with the existence of mathematical techniques for analyzing linear time-invariant systems, accounts for their central role in engineering.

For such a system the typical problem is to determine the response $x(t)$ to an excitation $f(t)$ applied (or turned on) at some instant of time, which is taken to be $t = 0$. When the input is applied, the system could be at rest (output initially zero) or there could be some output resulting from a previous excitation. With a circuit, for example, capacitors could still be discharging, and the output for $t > 0$ will then depend on the situation obtaining at $t = 0$ (i.e., on the *initial state* of the system).

Mathematical description

For signals that are continuous functions rather than specified at discrete intervals of time as with sampled data, a linear time-invariant system is represented by an ordinary linear differential equation with constant coefficients of the form

$$(a_0 D^n + a_1 D^{n-1} + \cdots + a_n)x(t) = f(t), \tag{3.1}$$

where $f(t)$ is the input, $x(t)$ is the output, and the differential operator describes the system. For any pair of functions satisfying this equation, it can be shown that the conditions of linearity and time-invariance are satisfied. The function $f(t)$ is, in general, piecewise-continuous, corresponding to an input which is turned on at $t = 0$, and the output $x(t)$ is then unique subject to the imposition of n initial values (or constraints) that define the system's initial state. When expressed in this manner, the determination of $x(t)$ for $t \geq 0$ is nothing more than the solution of an initial-value problem for the differential equation (3.1).

This formulation has some implications that are worth examining. Equation (3.1) says, in effect, that the input is obtained by applying a differential operator to the output, but since differentiation is the inverse of integration, a physically more reasonable statement is that the output is obtained by integrating the input. A system is therefore an *integrating device*, and an nth-order system represented by an nth-order differential equation performs an n-fold integration.

Integration is a smoothing or averaging process. The integral of a jagged function is smoother with more rounded peaks, and for a piecewise-continuous function, integration eliminates the jump discontinuities. As an example, if

$$f(t) = u(t),$$

then

$$D_0^{-1} f(t) = t u(t)$$

and

$$D_0^{-2} f(t) = \tfrac{1}{2} t^2 u(t).$$

These are plotted in Fig. 3.2, and whereas $f(t)$ is discontinuous at $t = 0$,

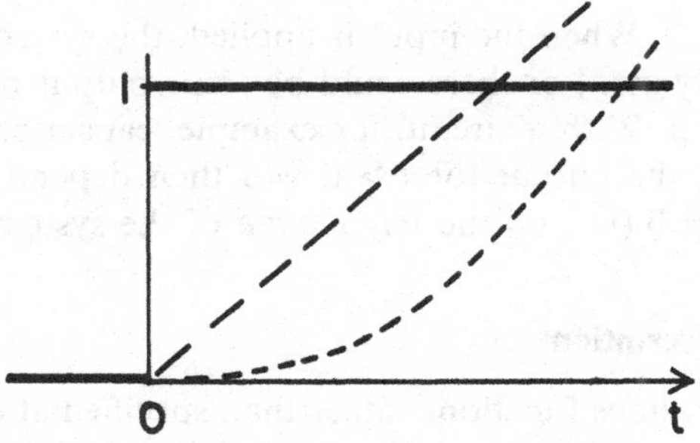

Fig. 3.2: Effect of integration applied to $u(t)$ (——); $D_0^{-1}u(t)$ (— —); $D_0^{-2}u(t)$ (---).

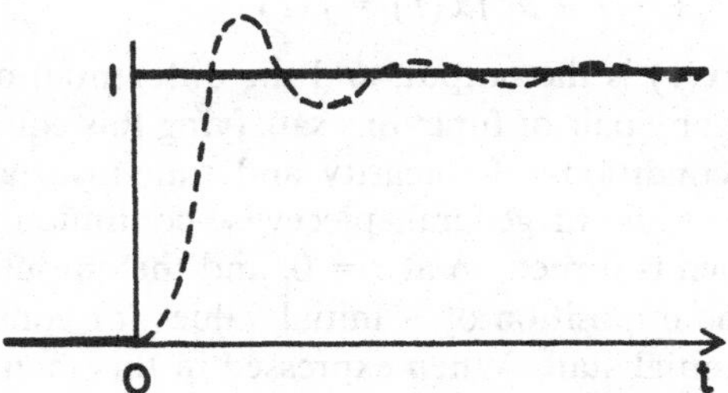

Fig. 3.3: Idealized (——) and more realistic (---) switch performance.

$D_0^{-1}f(t)$ and all subsequent integrals are continuous. It follows that even if the input $f(t)$ to a system has jump discontinuities, the output (assuming $n > 0$) and its first $n-1$ derivatives are all continuous. Indeed, any physically meaningful function must be continuous.

Why then do we allow functions with jump discontinuities as inputs to a system? In practice the input itself is the output of another system and must therefore be continuous. When a switch is closed to allow current to flow in a circuit, one could imagine that the current changes discontinuously from 0 to (let us say) 1, but a closer inspection shows that the actual current behaves as indicated in Fig. 3.3. After all, the switch is a mechanical system for which the input is the position of a control knob and the output is the position of a contact in a circuit, or more helpfully, the current itself; and even the position of the fingers (human or robotic) on the knob is the output of some additional system.

There are two reasons why certain types of discontinuous functions are of interest. In the first place they are convenient idealizations. In the preceding example the unit-step function represents an idealized behavior, which all switches approximate to some degree, and by using it as part of the input, the output is divorced from the idiosyncracies of any one switch. Secondly, the responses of a system to certain simple and, in general, discontinuous input completely characterize the system; and if measured or otherwise found, they provide all possible information about the system. One such input is the unit-step function itself.

3.2 Step response

This is occasionally referred to as the indicial response, but the term *step response* is more customary and descriptive. It is defined as follows.

> The step response is the output of a system initially at rest to which the input $u(t)$ is applied.

It is therefore the solution of

$$(a_0 D^n + a_1 D^{n-1} + \cdots + a_n)x(t) = f(t)$$

with

$$f(t) = u(t)$$

and all initial conditions zero.

We denote the step response by $g(t)$.

Example. Find the step response of the system defined by the differential equation

$$(D^2 + 3D + 2)x(t) = f(t).$$

In mathematical terms the task is to find the solution $g(t)$ of the equation

$$(D^2 + 3D + 2)g(t) = u(t)$$

with $g(0) = g'(0) = 0$. Obviously $g(t) = 0$ for $t < 0$ (see Section 2.10), and by applying a Laplace transform, we obtain

$$(s^2 + 3s + 2)G(s) = \frac{1}{s},$$

where $G(s) = \mathscr{L}\{g(t)\}$. Thus

$$G(s) = \frac{1}{s(s+1)(s+2)} = \frac{1/2}{s} - \frac{1}{s+1} + \frac{1/2}{s+2}$$

giving

$$g(t) = \tfrac{1}{2}(1 - 2e^{-t} + e^{-2t})u(t). \tag{3.2}$$

This is plotted in Fig. 3.4, along with the input $u(t)$, and we observe that

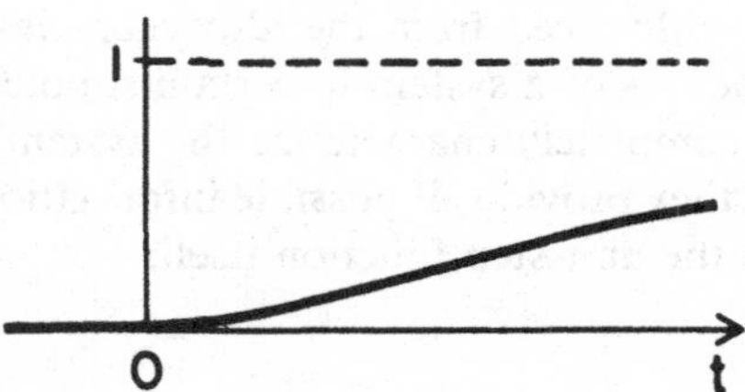

Fig. 3.4: The step response $g(t)$ of Eq. (3.2) compared with the input $u(t)$.

$g(t)$ and $g'(t)$ are continuous everywhere. Indeed, for an nth-order equation, $g(t), g'(t), \ldots, g^{(n-1)}(t)$ are all continuous.

The above example shows that

$$g(t) = \mathscr{L}^{-1}\left\{\frac{1}{sV(s)}\right\}, \tag{3.3}$$

where $V(s)$ is the characteristic function. It is therefore apparent that from a knowledge of $g(t)$ we can deduce $V(s)$ and, hence, the differential equation that describes the system.

Example. Given the step response $g(t) = (1 - \cos t)u(t)$, determine the differential equation that characterizes the system.

Since

$$G(s) = \frac{1}{s} - \frac{s}{s^2 + 1} = \frac{1}{s(s^2 + 1)},$$

it follows that

$$V(s) = s^2 + 1.$$

But $V(s)$ is the characteristic function obtained by replacing the differential operator D with s. The differential equation is therefore

$$(D^2 + 1)x(t) = f(t).$$

Admittedly this process would be difficult to carry out if the step response were known only as a measured curve, but the example demonstrates that the response $g(t)$ to a unit-step input contains the information necessary to characterize the system. We now introduce another input, which has the same effect.

3.3 Impulse function

Suppose you hit something – the desk, let us say – with your fist. The time that your fist is in contact with the desk may be very short indeed, but during that time the force applied builds up rapidly to a high value, then just as rapidly decreases to zero. By and large the effect on the desk is determined not so much by the duration of contact as by the momentum communicated, which is proportional to the time integral of the force. This type of situation comes up in mechanics and in other branches of science as well, and it is often convenient to consider an idealized extreme in which the force $f(t)$ is applied for only an infinitesimal amount of time but still communicates a nonzero amount of momentum.

The key properties of the resulting function are that $f(t)$ is nonzero for only an infinitesimal range of time, which we shall center on $t = 0$, and that $\int_{-\infty}^{\infty} f(t)\,dt$ has a finite nonzero value. By appropriate normalization this value can be taken equal to unity, and the properties then are

(i) $\quad f(t) = 0 \qquad \text{for } t \neq 0$

and

(ii) $\quad \int_{-\infty}^{\infty} f(t)\,dt = 1.$

It is obvious that the function is a very peculiar one indeed, and quite different from the rather ordinary ones discussed so far (and in this context even a piecewise-continuous function is ordinary). A crude approximation is provided by

$$f_\varepsilon(t) = \frac{1}{\varepsilon}\left\{u\left(t + \frac{\varepsilon}{2}\right) - u\left(t - \frac{\varepsilon}{2}\right)\right\}, \tag{3.4}$$

which is a rectangular pulse of height $1/\varepsilon$ and width ε centered on $t = 0$. For any $\varepsilon > 0$, $\int_{-\infty}^{\infty} f_\varepsilon(t)\,dt = 1$, and as ε is decreased the pulse height increases and the width decreases so as to maintain this value. Consider, then, a sequence of functions $f_\varepsilon(t)$ in which ε is progressively reduced, for example, by taking $\varepsilon = 1/n$ and allowing n to approach infinity through integer values. Three such functions having $\varepsilon = 1$, $\frac{1}{2}$, and $\frac{1}{3}$ are shown in Fig. 3.5, and for every $\varepsilon \neq 0$ $f_\varepsilon(t)$ is well defined. The limit of the sequence as $\varepsilon \to 0$ is a function having the properties (i) and (ii).

Other functions could be used to form the sequence. To avoid difficulties that might occur because of the piecewise continuity of the function in Eq. (3.4), we could employ instead a triangular pulse of height $1/\varepsilon$ and

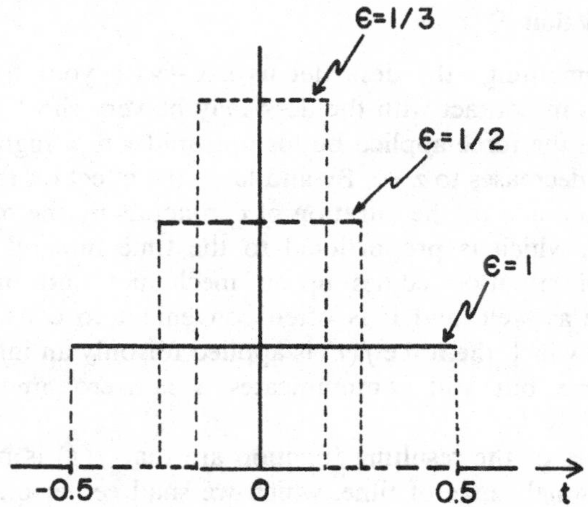

Fig. 3.5: The rectangular pulse function $f_\varepsilon(t)$ of Eq. (3.4) for $\varepsilon = 1$, $1/2$, and $1/3$.

width 2ε represented by

$$f_\varepsilon(t) = \frac{1}{\varepsilon^2}\{(t+\varepsilon)u(t+\varepsilon) - 2tu(t) + (t-\varepsilon)u(t-\varepsilon)\}, \tag{3.5}$$

which is a piecewise-smooth function for all $\varepsilon > 0$, or a Gaussian (bell-shaped) pulse of height $(\pi\varepsilon)^{-1/2}$ and half-width $\varepsilon^{1/2}$ represented by

$$f_\varepsilon(t) = (\pi\varepsilon)^{-1/2}e^{-t^2/\varepsilon}, \tag{3.6}$$

which is entirely smooth. This is plotted in Fig. 3.6 for $\varepsilon = 1/4, 1/20$, and $1/100$. Still others could suffice and as $\varepsilon \to 0$ all specify a limit function that is a particular example of a *generalized function*. The generalized function so defined is called the *impulse function* and is denoted by $\delta(t)$. An alternative name is the *Dirac delta function*, after Paul Adrien Maurice Dirac (1902–84), the English physicist who first proposed it, and thus

$$\delta(t) = \lim_{\varepsilon \to 0} f_\varepsilon(t), \tag{3.7}$$

where the limit is that of the sequence of functions given in Eq. (3.4), (3.5), or (3.6).

An impulse of strength a acting at time $t = 0$ is written as $a\delta(t)$, and is generally shown as a vertical arrow with a number beside it showing the strength (see Fig. 3.7). The physical interpretation depends on the

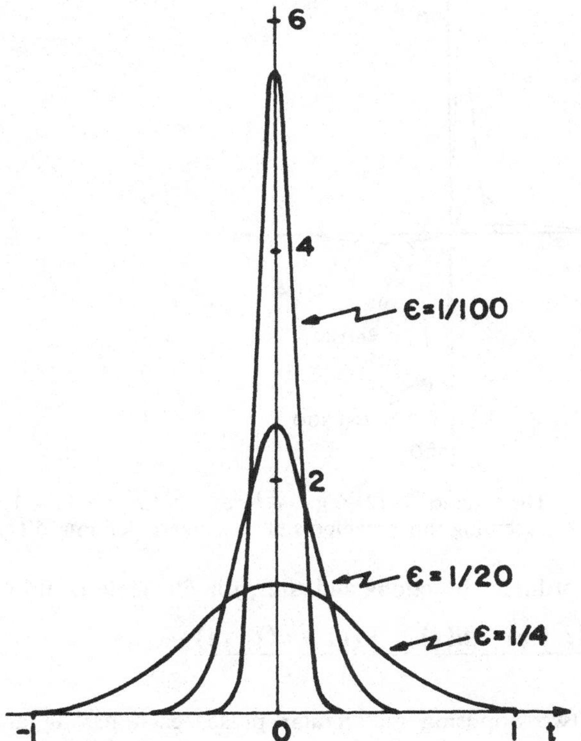

Fig. 3.6: The Gaussian pulse function $f_\varepsilon(t)$ of Eq. (3.6) for $\varepsilon = 1/4$, $1/20$, and $1/100$.

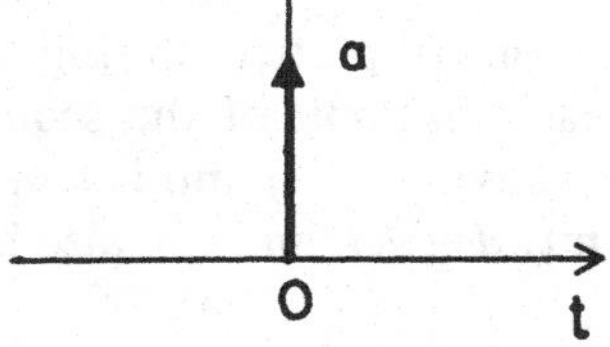

Fig. 3.7: Representation of an impulse of strength a at $t = 0$.

discipline. If, for example, $f(t)$ is a current in amperes with t measured in seconds, then $f(t) = a\delta(t)$ represents an impulsive current that transfers instantaneously at $t = 0$ a charge of a coulombs. In practice a very narrow pulse or spike whose time integral is a provides an approximate realization.

The impulse is not the only generalized function, and other examples are the derivatives of $\delta(t)$ (i.e., $\delta'(t)$, $\delta''(t)$, etc.). Each can be defined by

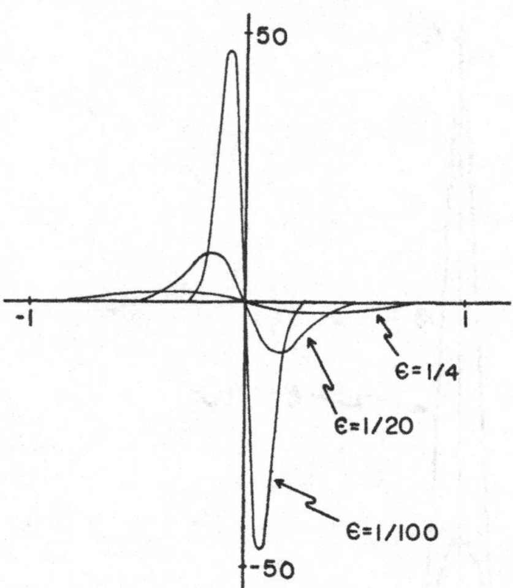

Fig. 3.8: The function $-(2t/\varepsilon)(\pi\varepsilon)^{-1/2}e^{-t^2/\varepsilon}$ for $\varepsilon = 1/4$, $1/20$, and $1/100$ representing three members of a sequence defining $\delta'(t)$.

a sequence of ordinary functions and thus, for $\delta'(t)$, we could consider

$$\frac{1}{\varepsilon}\left\{\frac{u(t+\varepsilon)-u(t)}{\varepsilon}-\frac{u(t)-u(t-\varepsilon)}{\varepsilon}\right\}$$

consisting of two opposing rectangular pulses each having area $1/\varepsilon$, spaced ε apart in time, or

$$-\frac{2t}{\varepsilon}(\pi\varepsilon)^{-1/2}e^{-t^2/\varepsilon}$$

obtained by differentiating Eq. (3.6). This is plotted in Fig. 3.8 for $\varepsilon = 1/4$, $1/20$, and $1/100$. In either case the limit of the sequence as $\varepsilon \to 0$ defines a function that is the derivative of $\delta(t)$ and is denoted by $\delta'(t)$. It is sometimes referred to as a (unit) *doublet*, and can also be given a physical interpretation.

The impulse function and, to a lesser degree, its derivatives, play a vital role in systems theory. None of them exists within the realm of classical mathematics, and because of this their application continues to produce controversy and occasional disagreements. Each can be defined as the limit of a sequence of ordinary functions or, more rigorously, using the theory of distributions developed in 1950 by the French mathematician Laurent Schwartz (1915–). Though the theory is beyond the scope of this text, it does constitute a resource to verify conclusions reached by reasoning less sound.

Properties of $\delta(t)$

A variety of results can be obtained using only the defining properties (i) and (ii), and these we present first. In all of the following, a is a real constant and $p(t)$ is assumed to be an ordinary function.

1. Property (i) is

$$\delta(t) = 0, \qquad \text{for } t \neq 0. \tag{3.8}$$

By a shift of variable we deduce

$$\delta(t-a) = 0, \qquad \text{for } t \neq a$$

and

$$p(t)\delta(t-a) = p(a)\delta(t-a) \tag{3.9}$$

provided $p(t)$ is continuous at $t = a$. If $p(t)$ has a jump discontinuity at $t = a$, for example, if $p(t) = u(t-a)$, the product is undefined. In the particular case when $p(a) = 0$ with $p'(a)$ finite,

$$p(t)\delta(t-a) = 0. \tag{3.10}$$

This is important for subsequent applications and amounts to a restriction on the order of infinity of $\delta(t)$ at $t = 0$.

2. Property (ii) is

$$\int_{-\infty}^{\infty} \delta(t)\,dt = 1. \tag{3.11}$$

From this and from (i) we deduce

$$\int_{-a}^{a} \delta(t)\,dt = 1$$

for any $a > 0$, and hence

$$\int_{-\infty}^{\infty} p(t)\delta(t)\,dt = \int_{-a}^{a} p(t)\delta(t)\,dt = p(0)$$

provided $p(t)$ is continuous at $t = 0$. Similarly, if $p(t)$ is continuous at $t = a$,

$$\int_{-\infty}^{\infty} p(t)\delta(t-a)\,dt = p(a), \tag{3.12}$$

which is the *sifting property* of $\delta(t)$.

3. $$\int_{-\infty}^{\infty} p(t)\delta'(t)\,dt = [p(t)\delta(t)]_{-\infty}^{\infty} - \int_{-\infty}^{\infty} p'(t)\delta(t)\,dt \qquad \text{by parts}$$

$$= -p'(0)$$

by the sifting property, provided $p'(t)$ is continuous at $t = 0$. Similarly,

$$\int_{-\infty}^{\infty} p(t)\delta'(t-a)\,dt = -p'(a)$$

for any finite a, and if a superscript in parentheses denotes the derivative,

$$\int_{-\infty}^{\infty} p(t)\delta^{(n)}(t-a)\,dt = (-1)^n p^{(n)}(a) \tag{3.13}$$

provided $p^{(n)}(t)$ is continuous at $t = a$.

4. From properties (i) and (ii)

$$\int_{-\infty}^{t} \delta(t')\,dt' = u(t), \tag{3.14}$$

and this is vital for the sequel.

For the most part we have so far managed to keep $\delta(t)$ and its derivatives underneath the integral sign. The big step is to spring $\delta(t)$ loose and connect it to a function we are already familiar with. This we do now.

5. From Eq. (3.14) by differentiation

$$\delta(t) = Du(t). \tag{3.15}$$

This gives a meaning to the derivative of a piecewise-continuous function at a point of jump discontinuity, and though it is only a formal relation, it is important for systems applications. We deduce

$$\delta(t-a) = Du(t-a)$$

and

$$\delta'(t-a) = D^2u(t-a), \quad \text{and so forth.}$$

As previously remarked, the derivatives of $\delta(t)$ are also generalized functions.

Laplace transform of $\delta(t)$

Since the application of a Laplace transform was previously limited to functions that were piecewise-continuous, it is not surprising that the transformation of the generalized function $\delta(t)$ requires some care. There are several ways to obtain the transform, and three of them are presented here.

Consider $\delta(t-a)$ where $a > 0$. From the definition of a Laplace transform

$$\mathscr{L}\{\delta(t-a)\} = \int_0^\infty e^{-st}\delta(t-a)\,dt$$
$$= e^{-sa}$$

by using the sifting property of Eq. (3.12). Similarly, from Eq. (3.13),

$$\mathscr{L}\{\delta^{(n)}(t-a)\} = (-1)^n \frac{d^n}{dt^n} e^{-st}\bigg|_{t=a} = s^n e^{-sa},$$

and in the limit as $a \to 0+$ we have

$$\mathscr{L}\{\delta^{(n)}(t)\} = s^n \tag{3.16}$$

for $n = 0, 1, 2, \ldots$. This is sufficiently important to be treated as a memorable transform and inserted as the fifth (and final) entry in the table of memorable transforms in Section 2.4. In particular, for $n = 0$

$$\mathscr{L}\{\delta(t)\} = 1. \tag{3.17}$$

Alternatively, using the derivative property of the Laplace transform in conjunction with Eq. (3.15),

$$\mathscr{L}\{\delta(t-a)\} = \mathscr{L}\{Du(t-a)\}$$
$$= s\mathscr{L}\{u(t-a)\} - 0$$
$$= e^{-sa},$$

and in the limit as $a \to 0+$ we again obtain Eq. (3.17).

The initial displacement of the impulse function is necessary to ensure that it lies entirely in the range $t > 0$ and can be regarded as zero for $t < 0$. Failure to do so would invalidate the use of Eq. (3.12) and the derivative property, as indicated by the following:

$$\mathscr{L}\{\delta(t)\} = \mathscr{L}\{Du(t)\}$$
$$= s\mathscr{L}\{u(t)\} - u(0+)$$
$$= s\frac{1}{s} - 1 = 0,$$

which is obviously in error.

The third method harks back to the definition of $\delta(t)$ using a sequence of ordinary functions each of which can be transformed by the methods originally established. In particular, consider Eq. (3.7) with $f_\varepsilon(t)$ defined as a rectangular pulse that is shifted into the region $t > 0$ by writing

$$f_\varepsilon(t) = \frac{1}{\varepsilon}\{u(t) - u(t-\varepsilon)\}.$$

Then

$$\mathscr{L}\{f_\varepsilon(t)\} = \frac{1}{s\varepsilon}(1 - e^{-s\varepsilon})$$

and

$$\begin{aligned}\mathscr{L}\{\delta(t)\} &= \int_0^\infty e^{-st}\Big\{\lim_{\varepsilon\to 0} f_\varepsilon(t)\Big\}\,dt\\ &= \lim_{\varepsilon\to 0}\frac{1}{s\varepsilon}(1 - e^{-s\varepsilon})\\ &= 1\end{aligned}$$

using the Taylor-series expansion of the exponential or l'Hospital's rule. Though this derivation is simple and straightforward, it is not as rigorous as it seems at first sight. In proceeding from the first to the second lines, we have assumed that the integral of a limit is the limit of an integral. This is true only for a uniformly convergent sequence, which is not the case here.

Indeed, all three methods are open to criticism (e.g., the validity of the derivative property has not been established for generalized functions), and it is therefore comforting to know that the same results are produced by the theory of distributions.

3.4 Equivalent input

We now return to the Laplace-transform solution of a differential equation. As shown in Section 2.10, the Laplace transform $X(s)$ of the output $x(t)$ of an nth-order system represented by an nth-order differential equation is

$$X(s) = \frac{1}{V(s)}\{A(s) + F(s)\}, \tag{3.18}$$

where $V(s)$ is the characteristic function, $F(s)$ is the transform of the input $f(t)$, and $A(s)$ is a polynomial in s of degree less than or equal to $n - 1$ produced by the initial values. Thus

$$A(s) = c_0 + c_1 s + \cdots + c_{n-1}s^{n-1}, \tag{3.19}$$

where the cs are constants. If the system is initially at rest (initial values all zero),

$$c_0 = c_1 = \cdots = c_{n-1} = 0,$$

implying $A(s) = 0$, and in this case

$$x(t) = \mathscr{L}^{-1}\left\{\frac{F(s)}{V(s)}\right\}. \tag{3.20}$$

The *equivalent input* (or excitation) $\tilde{f}(t)$ is that input that when applied to the system at rest, yields the same output as in the given problem. If $\tilde{F}(s) = \mathscr{L}\{\tilde{f}(t)\}$ the requirement is

$$X(s) = \frac{\tilde{F}(s)}{V(s)}$$

and comparison with Eq. (3.18) now shows

$$\tilde{F}(s) = F(s) + A(s).$$

The equivalent input is therefore

$$\tilde{f}(t) = f(t) + \mathscr{L}^{-1}\{A(s)\}, \tag{3.21}$$

that is,

$$\tilde{f}(t) = f(t) + c_0\delta(t) + c_1\delta'(t) + \cdots + c_{n-1}\delta^{(n-1)}(t),$$

where the constants depend on the initial values. This is one reason for the occurrence of generalized functions in the input to a system.

The advantage gained by using the equivalent input is not in the shortening of the process for computing a solution, but rather in the conceptual simplification of the problem. Since the initial conditions are absorbed into the input, solutions can be obtained for all possible initial values with the system treated as if it were at rest. We emphasize that the system is not actually at rest initially, nor (of course) does the solution show it to be, but the analysis is carried out as if it were.

Example. For the initial-value problem $(D^3 + D)x(t) = (\sin t)u(t)$ with $x(0) = 1$, $x'(0) = 3$, and $x''(0) = 4$, find the equivalent input $\tilde{f}(t)$.

By application of a Laplace transform to the left-hand side of the equation, we obtain

$$\begin{aligned}\mathscr{L}\{(D^3 + D)x(t)\} &= s^3X(s) - s^2(1) - s(3) - 4 + sX(s) - 1\\ &= (s^3 + s)X(s) - (5 + 3s + s^2).\end{aligned}$$

Hence

$$A(s) = 5 + 3s + s^2 = \mathscr{L}\{5\delta(t) + 3\delta'(t) + \delta''(t)\}$$

and

$$\tilde{f}(t) = (\sin t)u(t) + 5\delta(t) + 3\delta'(t) + \delta''(t).$$

For any given problem, the use of an equivalent input does not materially assist in finding the output, and as the above example shows, if the order of the system is greater than unity, it is necessary to invoke the Laplace-transform technique to find $\tilde{f}(t)$. Nevertheless, the equivalent input is a worthwhile concept and we shall use it later.

Exercises

1. Using Eq. (3.15), show that $\delta(t) = \delta(-t)$ and deduce the analogous results for $\delta'(t)$ and $\delta''(t)$.

2. Find the first and second derivatives of:
*(i) $(t+1)u(t)$. (ii) $t\{1 - u(t)\} + t^2 u(t)$.
(iii) $(\sin t)\{u(t - \pi/2) - u(t - \pi)\}$.
(iv) $|t|$ [*Hint*: Express $|t|$ in terms of $u(t)$].

3. Evaluate the following integrals:
*(i) $\int_{-\infty}^{\infty} (t+1)^2 \delta(t)\, dt$. (ii) $\int_0^{\infty} t^2 \delta(t-1)\, dt$.
(iii) $\int_{-\infty}^{\infty} \{e^t - (e^t + e^{-t})u(t)\}\, dt$.
(iv) $\int_0^{\infty} e^{-t}\delta'(t-1)\, dt$.
(v) $\int_{-\infty}^{\infty} te^{-t}\{u(t) + 2\delta'(t) + \delta''(t)\}\, dt$.
(vi) $\int_0^{\infty} t^2\{\delta'(t-1) + 2\delta''(t-2)\}\, dt$.

4. If $\delta(t)$ is defined as the limit of a sequence of triangular pulse functions $f_\varepsilon(t)$ similar (but not identical) to that given in Eq. (3.5), find $\mathscr{L}\{f_\varepsilon(t)\}$ and, hence, show $\mathscr{L}\{\delta(t)\} = 1$.

*5. Given the system defined by the differential equation $(D^2 + 2D)x(t) = f(t)$ with $f(t) = e^{-t}u(t)$ and $x(0) = 1$, $x'(0) = 2$, find the equivalent input $\tilde{f}(t)$.

6. Given the system defined by the differential equation $(D^2 + 1)x(t) = f(t)$, find the output $x(t)$ when the equivalent input is $\tilde{f}(t) = tu(t) + 2\delta(t) + \delta'(t)$.

7. (i) If $f(t) = (t+1)^2 u(t)$, find $D^2 f(t)$.
(ii) Using your answer to (i) and the TMT, find $\mathscr{L}\{D^2 f(t)\}$.
(iii) The derivative property of the Laplace transform enables you to express $\mathscr{L}\{D^2 f(t)\}$ in terms of $f(0)$ and $f'(0)$. What does your answer to (ii) imply as regards $f(0)$ and $f'(0)$?
(iv) Compare the value of $f(0)$ obtained in (iii) with that found from the expression for $f(t)$. If they disagree, explain why.

8. Repeat the preceding problem using $f(t) = (t+1)^2 u(t - a)$ with $a > 0$.

3.5 Impulse response

We previously introduced the step response and noted that it provides a complete specification of a system. Even more important is the response based on the *impulse function* as input, and its definition parallels that of the step response.

> The impulse response is the output of a system initially at rest to which the input $\delta(t)$ is applied.

It is therefore the solution of

$$(a_0 D^n + a_1 D^{n-1} + \cdots + a_n)x(t) = f(t)$$

with

$$f(t) = \delta(t)$$

and all initial conditions zero. We denote the impulse response by $h(t)$.

It is of interest (but no practical significance) that for the nth-order system given above $h(t)$ is the same for $t > 0$ as if the input were zero with $h(0) = h'(0) = \cdots = h^{(n-2)}(0) = 0$ but $h^{(n-1)}(0) = 1/a_0$.

Example. Given the system defined by the differential equation $(D^2 + 3D + 2)x(t) = f(t)$, find the impulse response.

The task is to solve the equation

$$(D^2 + 3D + 2)h(t) = \delta(t)$$

with $h(0) = h'(0) = 0$. Application of a Laplace transform gives

$$(s^2 + 3s + 2)H(s) = 1,$$

where $H(s) = \mathscr{L}\{h(t)\}$. Hence

$$H(s) = \frac{1}{(s+1)(s+2)} = \frac{1}{s+1} - \frac{1}{s+2},$$

from which we have

$$h(t) = (e^{-t} - e^{-2t})u(t).$$

We observe that $h(t)$ is continuous everywhere and, indeed, for an nth-order system, $h(t), h'(t), \ldots, h^{(n-2)}(t)$ are all continuous.

From this example or from the definition of the impulse response it is evident that

$$\mathscr{L}\{h(t)\} = H(s) = \frac{1}{V(s)}. \tag{3.22}$$

$H(s)$ is called the *transfer function* and is the basis for the s-plane analysis of systems. In terms of $H(s)$ the output $x(t)$ of any system is, from Eq. (3.20),

$$x(t) = \mathscr{L}^{-1}\{H(s)\tilde{F}(s)\}, \tag{3.23}$$

where $\tilde{F}(s)$ is the transform of the equivalent input. For a system that is initially at rest, $\tilde{F}(s) = F(s)$.

Since $\delta(t) = Du(t)$, it is natural to expect a simple connection between the impulse and step responses. From Eqs. (3.3) and (3.22) we have

$$G(s) = \frac{H(s)}{s}, \tag{3.24}$$

where $G(s)$ is the Laplace transform of the step response $g(t)$. Hence

$$\begin{aligned}\mathscr{L}\{Dg(t)\} &= sG(s) - g(0)\\ &= H(s) = \mathscr{L}\{h(t)\}\end{aligned}$$

since $g(0) = 0$ for a system of nonzero order. It follows that

$$h(t) = Dg(t) \tag{3.25}$$

and conversely

$$g(t) = D_0^{-1}h(t). \tag{3.26}$$

We can therefore derive the impulse response from the step response and vice versa. As an example, if

$$g(t) = \tfrac{1}{2}(1 - 2e^{-t} + e^{-2t})u(t),$$

then

$$\begin{aligned}h(t) &= D\{\tfrac{1}{2}(1 - 2e^{-t} + e^{-2t})u(t)\}\\ &= D\{\tfrac{1}{2}(1 - 2e^{-t} + e^{-2t})\}u(t) + \tfrac{1}{2}(1 - 2e^{-t} + e^{-2t})Du(t)\\ &= (e^{-t} - e^{-2t})u(t),\end{aligned}$$

since the quantity multiplying $Du(t) = \delta(t)$ vanishes at $t = 0$.

The impulse response and its transform provide two additional ways to characterize a system, and the methods that are now available can be

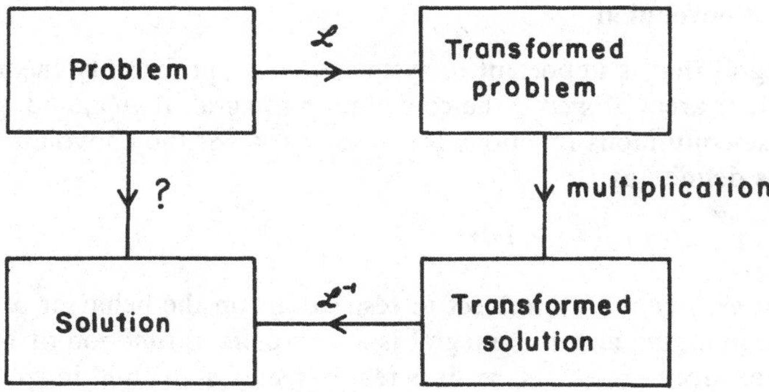

Fig. 3.9: Block diagram of the Laplace-transform method of solution of a system problem.

illustrated as follows:

$$\text{differential equation} \leftrightarrow \text{characteristic function } V(s) = \frac{1}{H(s)},$$

$$\text{impulse response } h(t) \leftrightarrow \text{transfer function } H(s),$$

$$\text{step response } g(t) \leftrightarrow G(s) = \frac{H(s)}{s}.$$

The descriptions on the left are in the t plane, and any one completely specifies the system. In contrast to the differential equation, the impulse and step responses are, in principle, measurable quantities, and enable the system to be determined using measurements alone. The descriptions on the right are in the s plane, and all are related to the transfer function, which is the vital ingredient of a Laplace-transform solution of a systems problem.

When a system is excited by an input, the key steps in the determination of the output are evident from Eq. (3.23). The analysis is carried out in the s plane and the first step is to find the Laplace transform of the equivalent input $\tilde{f}(t)$ or, if the system is initially at rest, of the input $f(t)$ alone. The resulting transform is then multiplied by the transfer function $H(s)$, and finally the output is obtained by application of an inverse transform tantamount to a process of recognition. The steps are illustrated in Fig. 3.9, and the diagram suggests that there ought to be a way to solve the problem directly without recourse to a Laplace transform. We now consider this.

3.6 Convolution

An integral that is important in systems theory, probability theory, and many other areas as well is the convolution integral. If $p(t)$ and $q(t)$ are piecewise-continuous functions for $-\infty < t < \infty$, the convolution of p and q is defined as

$$\int_{-\infty}^{\infty} p(\tau)q(t-\tau)\,d\tau$$

and written $p(t) * q(t)$. Subject to restrictions on the behavior of p and q for large arguments, the integral is a well-defined function of t that is continuous for $-\infty < t < \infty$. It is real if p and q are, and in spite of its appearance, it is symmetric in p and q. This can be seen by writing $\tau' = t - \tau$, and thus

$$p(t) * q(t) = \int_{-\infty}^{\infty} p(\tau)q(t-\tau)\,d\tau = \int_{-\infty}^{\infty} p(t-\tau)q(\tau)\,d\tau. \tag{3.27}$$

The integrand entails the folding over of one function and its displacement by an amount t relative to the other; some older texts refer to the convolution integral as the *Faltung integral* from the German word for folding. To appreciate this action, consider the functions $p(t)$ and $q(t)$ shown in the upper part (a) of Fig. 3.10. For a particular t, $p(\tau)$ and $q(t-\tau)$ are shown in part (b) and the product in part (c). The shaded area under the curve is the value of the convolution for this particular t. Repetition of the process for other t then leads to the determination of the convolution for all t, $-\infty < t < \infty$, though it is obvious that this could be extremely time-consuming to carry out.

The notable exception is when one or the other of the functions $p(t)$ and $q(t)$ is the impulse function or a derivative thereof (the product of two generalized functions is undefined). Justification for applying a convolution in this case is based on the arguments advanced in Section 3.3, and if $p(t) = \delta^{(n)}(t)$, the sifting property gives

$$\delta^{(n)}(t) * q(t) = q^{(n)}(t).$$

In particular

$$\delta(t) * q(t) = q(t)$$

showing that convolution of a function with $\delta(t)$ reproduces the function.

The integral simplifies somewhat if $p(t)$ and $q(t)$ are zero for $t < 0$. Because $p(\tau) = 0$ for $\tau < 0$,

$$p(t) * q(t) = \int_{0}^{\infty} p(\tau)q(t-\tau)\,d\tau \tag{3.28}$$

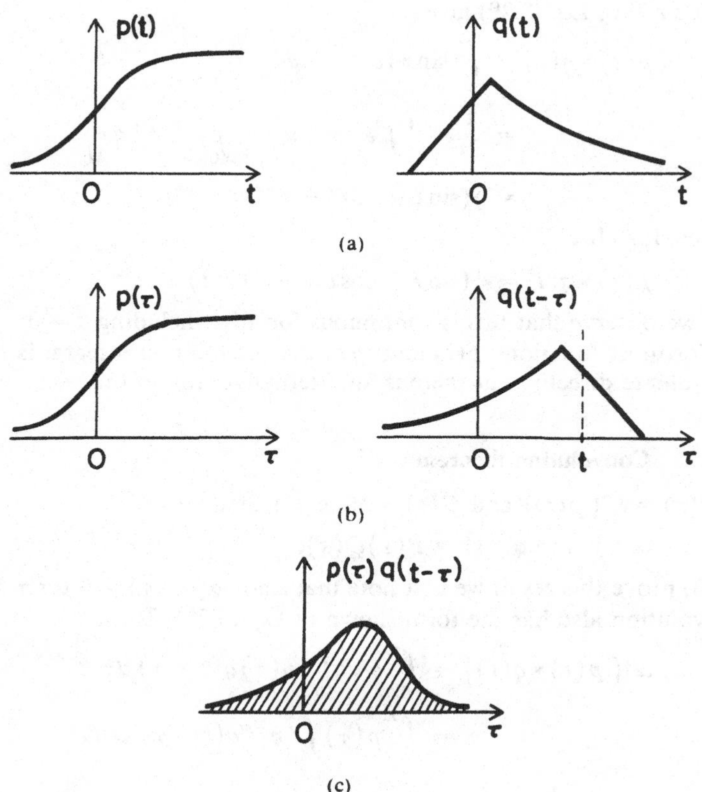

Fig. 3.10: Illustration of how a convolution is formed.

and since $g(t-\tau) = 0$ for $\tau > t$,

$$p(t) * q(t) = \int_0^t p(\tau) q(t-\tau)\, d\tau, \tag{3.29}$$

which can also be written as

$$p(t) * q(t) = \int_0^t p(t-\tau) q(\tau)\, d\tau. \tag{3.30}$$

Equation (3.29) or (3.30) is the convolution integral of interest in the context of Laplace transforms and is the form considered throughout the rest of this chapter. It is evident from these equations that

$$p(t) * q(t) = 0, \qquad \text{for } t < 0. \tag{3.31}$$

Example. Evaluate $p(t) * q(t)$ for $p(t) = (\sin t)u(t)$ and $q(t) = e^{-t}u(t)$.

For $t > 0$, Eq. (3.29) gives

$$p(t) * q(t) = \int_0^t (\sin \tau) e^{-(t-\tau)}\, d\tau$$

$$= \frac{1}{2i} e^{-t} \left\{ \int_0^t e^{(1+i)\tau}\, d\tau - \int_0^t e^{(1-i)\tau}\, d\tau \right\}$$

$$= \tfrac{1}{2}(\sin t - \cos t + e^{-t}).$$

Hence, for all t

$$p(t) * q(t) = \tfrac{1}{2}(\sin t - \cos t + e^{-t})u(t)$$

and we observe that this is continuous for all t including $t = 0$.

For most functions $p(t)$ and $q(t)$ the convolution integral is difficult to evaluate directly, but there is an alternative way to proceed.

Convolution theorem

If $P(s) = \mathscr{L}\{p(t)\}$ and $Q(s) = \mathscr{L}\{q(t)\}$ then

$$\mathscr{L}\{p(t) * q(t)\} = P(s)Q(s).$$

To prove this result we first note that since $q(t-\tau) = 0$ for $\tau > t$, the convolution also has the form shown in Eq. (3.28). Thus

$$\mathscr{L}\{p(t) * q(t)\} = \int_0^\infty e^{-st} \int_0^\infty p(\tau) q(t-\tau)\, d\tau\, dt$$

$$= \int_0^\infty p(\tau) \int_0^\infty e^{-st} q(t-\tau)\, dt\, d\tau$$

$$= \int_0^\infty p(\tau) e^{-s\tau}\, d\tau\, Q(s)$$

giving

$$\mathscr{L}\{p(t) * q(t)\} = P(s)Q(s), \tag{3.32}$$

where we have used the ability to reverse the order of integration for uniformly convergent integrals and the t-plane shift property of the Laplace transform. This is the *convolution theorem*. It states that the Laplace transform of a convolution is the product of the transforms of the functions involved, and is otherwise known as *Borel's theorem*, after the French mathematician Felix Borel (1871–1956). It also implies

$$p(t) * q(t) = \mathscr{L}^{-1}\{P(s)Q(s)\}. \tag{3.33}$$

The theorem can therefore be used to evaluate a convolution by conversion to an inverse Laplace transform, or to evaluate an inverse Laplace transform by conversion to a convolution.

Example 1. Use the convolution theorem to evaluate

$$\mathscr{L}^{-1}\left\{\frac{1}{(s^2+1)^2}\right\}.$$

Let

$$P(s) = Q(s) = \frac{1}{s^2+1}.$$

From the TMT we then have

$$p(t) = q(t) = (\sin t)u(t)$$

and hence, for $t > 0$,

$$\begin{aligned}\mathscr{L}^{-1}\left\{\frac{1}{(s^2+1)^2}\right\} &= \int_0^t \sin\tau \sin(t-\tau)\,d\tau \\ &= \sin t\int_0^t \sin\tau\cos\tau\,d\tau - \cos t\int_0^t \sin^2\tau\,d\tau \\ &= \tfrac{1}{4}\{\sin t(1-\cos 2t) - \cos t(2t - \sin 2t)\} \\ &= \tfrac{1}{2}(\sin t - t\cos t).\end{aligned}$$

It follows that

$$\mathscr{L}^{-1}\left\{\frac{1}{(s^2+1)^2}\right\} = \frac{1}{2}(\sin t - t\cos t)u(t).$$

Example 2. Use the convolution theorem to evaluate

$$\{t^2u(t)\} * \{(\cos 2t)u(t)\}.$$

If $p(t) = t^2u(t)$ and $q(t) = (\cos 2t)u(t)$, then

$$P(s) = \frac{2}{s^3} \quad\text{and}\quad Q(s) = \frac{s}{s^2+4}$$

giving

$$P(s)Q(s) = \frac{2}{s^2(s^2+4)} = \frac{1}{2}\left(\frac{1}{s^2} - \frac{1}{s^2+4}\right).$$

Hence

$$\begin{aligned}\{t^2u(t)\} * \{(\cos 2t)u(t)\} &= \mathscr{L}^{-1}\left\{\frac{1}{2}\left(\frac{1}{s^2} - \frac{1}{s^2+4}\right)\right\} \\ &= \tfrac{1}{4}(2t - \sin 2t)u(t).\end{aligned}$$

Based on the first of these examples, it might appear that the convolution theorem provides an effective method for evaluating an inverse Laplace transform, but the cases in which this is so are the exception rather than the rule. The direct evaluation of a convolution is usually quite tedious and the theorem is more often employed to facilitate the evaluation by conversion to an inverse Laplace-transform problem.

Systems application

It was shown in Section 3.5 that the output of a linear system is

$$x(t) = \mathscr{L}^{-1}\{H(s)\tilde{F}(s)\},$$

where $H(s)$ is the transfer function ($= \mathscr{L}\{h(t)\}$) and $\tilde{F}(s)$ is the transform of the equivalent input $\tilde{f}(t)$. The convolution theorem makes it possible to express this in a form that does not involve Laplace transforms, and using Eq. (3.33) we have

$$x(t) = h(t) * \tilde{f}(t) \tag{3.34}$$

for $t > 0$. The output is therefore the convolution of the impulse response of the system and the equivalent input, and can be computed by direct evaluation of the convolution integral, namely,

$$x(t) = \int_0^t h(\tau)\tilde{f}(t-\tau)\,d\tau = \int_0^t h(t-\tau)\tilde{f}(\tau)\,d\tau. \tag{3.35}$$

In the particular case of a system initially at rest, the equivalent input is the true input $f(t)$ and

$$x(t) = \int_0^t h(\tau)f(t-\tau)\,d\tau = \int_0^t h(t-\tau)f(\tau)\,d\tau. \tag{3.36}$$

If $f(t) = 0$ for $t < 0$, Eq. (3.36) is valid for all t.

Most inputs of practical significance are piecewise-continuous functions at worst and do not contain any generalized functions, but if the system is not initially at rest, generalized functions must occur as part of the equivalent input $\tilde{f}(t)$. Some care is then necessary when interpreting the resulting convolution. To see how to proceed, suppose

$$\tilde{f}(t) = f(t) + a\delta(t),$$

where $f(t)$ is assumed zero for $t < 0$. Since $h(t)$ is also zero for $t < 0$, it follows that

$$\begin{aligned}\int_0^t h(\tau)\tilde{f}(t-\tau)\,d\tau &= \int_{-\infty}^{\infty} h(\tau)\{f(t-\tau) + a\delta(t-\tau)\}\,d\tau \\ &= ah(t) + \int_0^t h(\tau)f(t-\tau)\,d\tau\end{aligned}$$

as implied by the convolution originally defined in Eq. (3.27). In effect, this requires the interpretation of the convolution integral as

$$\int_0^t h(\tau)\tilde{f}(t-\tau)\,d\tau = \lim_{\varepsilon\to 0+}\int_0^{t+\varepsilon} h(\tau)\tilde{f}(t-\tau)\,d\tau.$$

Similarly

$$\int_0^t h(t-\tau)\tilde{f}(\tau)\,d\tau = \lim_{\varepsilon\to 0+}\int_{-\varepsilon}^{t} h(t-\tau)f(\tau)\,d\tau$$

and such an infinitesimal extension of the range of integration is necessary whenever the system is not at rest or, more generally, whenever the input includes a generalized function acting at $t = 0$.

The importance of the impulse response is evident from Eqs. (3.35) and (3.36). If the impulse response of a system is known, the output is given explicitly as a convolution, and it is no longer necessary to invoke the differential equation defining the system. In principle at least, we can now separate the study of linear systems from any consideration of differential equations, but this type of extension of the system concept is beyond the scope of our treatment.

The output of a linear system can also be expressed in terms of the step response $g(t)$. Since

$$\mathscr{L}\{g(t)\} = G(s) = \frac{H(s)}{s},$$

it follows that

$$\mathscr{L}\{D_0^{-1}x(t)\} = \frac{X(s)}{s} = G(s)\tilde{F}(s) = \mathscr{L}\{g(t)*\tilde{f}(t)\}$$

giving

$$D_0^{-1}x(t) = g(t)*\tilde{f}(t),$$

and hence

$$x(t) = D\int_0^t g(\tau)\tilde{f}(t-\tau)\,d\tau = D\int_0^t g(t-\tau)\tilde{f}(\tau)\,d\tau. \qquad (3.37)$$

The inclusion of generalized functions as part of f requires the extension of the range of integration noted above, and it is then preferable to evaluate the integral before carrying out the derivative operation.

If the system is initially at rest and the input is at worst piecewise-continuous, the differentiation can be performed analytically. From the first expression shown in Eq. (3.37) we have

$$x(t) = f(0)g(t) + \int_0^t g(\tau)\frac{\partial}{\partial t}f(t-\tau)\,d\tau, \qquad (3.38)$$

which is usually referred to as *Duhamel's integral* after the French mathematician Jean Marie Constant Duhamel (1797–1872). Similarly, from the second expression in Eq. (3.37),

$$x(t) = \int_0^t \frac{\partial}{\partial t} g(t-\tau) f(\tau)\, d\tau, \tag{3.39}$$

where we have used the fact that for a system of nonzero order $g(0) = 0$. This follows directly from Eq. (3.36) by employing Eq. (3.25).

A knowledge of the step response is therefore sufficient to obtain the output of a system for any given input. Theoretically, at least, it is unnecessary to employ the Laplace-transform technique at all, though it may be expedient to do so to simplify the evaluation of the convolution integrals. We also remark that the expressions for the output $x(t)$ are somewhat heavier than those based on the impulse response $h(t)$, and this is one reason the description of a system by its impulse response is more convenient.

Exercises

1. (i) Use the convolution theorem to show that

$$\mathscr{L}^{-1}\left\{\frac{s}{(s^2+a^2)^2}\right\} = \frac{t}{2a}(\sin at)u(t).$$

   (ii) Solve the equation $(D^2+9)x(t) = 6(\cos 3t)u(t)$ with $x(0) = x'(0) = 0$.

*2. If the input $f(t) = (4e^{-t} + 8e^{-3t})u(t)$ is applied to a system at rest whose impulse response is $h(t) = \frac{1}{2}(e^{-t} - e^{-3t})u(t)$, use Eq. (3.36) to find the output.

3. If the input $f(t) = e^{-2t}(\cos t)u(t)$ is applied to a system at rest whose impulse response is $h(t) = te^{-2t}u(t)$, use the Laplace-transform technique to find the output.

4. If the step response of a system is $g(t) = \frac{1}{2}(\cosh t + \cos t - 2)u(t)$:
   (i) Find the transfer function $H(s)$.
   (ii) Determine the output when the input $f(t) = (\cos t)u(t)$ is applied to the system at rest.

5. If the input $f(t) = 2e^{-2t}$ is applied to a system at rest whose step response is

$$g(t) = \tfrac{1}{6}(2 - 3e^{-t} + e^{-3t})u(t):$$

   (i) Deduce the corresponding initial-value problem and then solve the differential equation for $t > 0$.

(ii) Compare the expression for $x(t)$ with that obtained from Eq. (3.38).

6. For some applications it is convenient to employ the *ramp response* $r(t)$ defined as the output of a system initially at rest to which the input $tu(t)$ is applied.
   (i) Show that

$$\mathscr{L}\{r(t)\} = R(s) = \frac{H(s)}{s^2}.$$

   (ii) If $r(t) = \frac{1}{8}(t^2 - \sin^2 t)u(t)$, find the differential equation defining the system.

7. If $h(t) = \frac{1}{2}\{1 + e^t(\sin t - \cos t)\}u(t)$ and $\tilde{f}(t) = 2\delta(t) - \delta'(t) - \delta''(t)$, use Eq. (3.35) to find $x(t)$ for $t > 0$.

8. If the equivalent input $\tilde{f}(t) = u(t) + \delta(t) - \delta'(t)$ is applied to a system whose transfer function is

$$H(s) = \frac{1}{s^2 + 1}:$$

   (i) Determine $h(t)$ and, hence, use Eq. (3.36) to find $x(t)$ for $t > 0$.
   (ii) Deduce a corresponding initial-value problem and then solve for $t > 0$.

*9. Solve, for $t > 0$, the integrodifferential equation

$$(D + 1)x(t) + \int_0^t x(\tau)e^{\tau - t}\, d\tau = 0$$

with $x(0) = 1$.

10. Solve, for $t > 0$, the integrodifferential equation

$$Dx(t) + 5\int_0^t (\cos 2\tau)x(t - \tau)\, d\tau = 27u(t)$$

with $x(0) = 9$.

3.7 Systems analysis

We now apply the procedures that have been developed to a general analysis of linear systems. To simplify the discussion it will be assumed that the input is applied at time $t = 0$ to a system at rest.

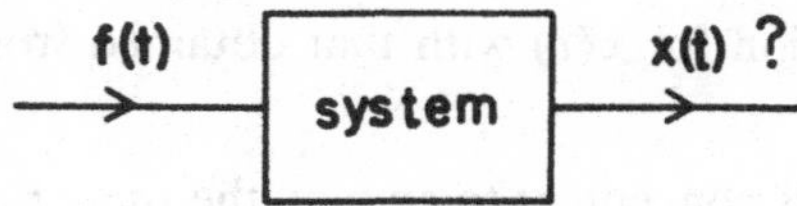

Fig. 3.11: The direct problem.

There are two types of system problems that can be addressed. The direct or standard one is that of finding the output of a system which is excited by a known input. This is illustrated in Fig. 3.11, where the question mark indicates the quantity to be found. Mathematically the task is equivalent to the solution of a differential equation, but the output is also given explicitly as a convolution, namely,

$$x(t) = h(t) * f(t), \tag{3.40}$$

from which $x(t)$ follows immediately if the impulse response is known. We remark that if the system is described by its differential equation, the determination of $h(t)$ is synonymous with solving the equation for a specific input. In practice the evaluation of the convolution integral is avoided by using the convolution theorem, in which case

$$x(t) = \mathscr{L}^{-1}\{H(s)F(s)\},$$

and since the transfer function can be obtained from either the impulse response or the differential equation, both methods come together in this approach.

There is also an indirect or inverse type of system problem, and we shall examine this first.

Deconvolution

For a small but growing number of applications the task is *not* to find the output knowing the system and the input, but to derive the input or a system description from a knowledge of the output and of the system or the input. In inverse-scattering and remote-sensing studies the transmitted and received signals are known and/or can be measured, and if these are regarded as the input and output, respectively, of a linear system, the task is to determine the object (or system) which has produced the output. In effect we are now seeking the impulse response or transfer function, and the question mark in Fig. 3.11 has been shifted to the box.

There are other applications where the input is sought. Any sensor or measurement device is itself a system, and as we pointed out in Section 3.1, a system has a smoothing or averaging effect which suppresses some

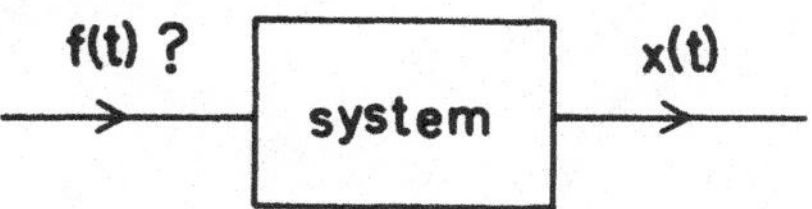

Fig. 3.12: The indirect problem.

of the sharpness and definition of the input. Since the input is the quantity which is actually desired, the task is to recover this from a knowledge of the output (see Fig. 3.12). It is possible in theory and to some extent in practice. An example is the deblurring of photographic images. The system in this case is a camera which, regardless of its sophistication, is not perfect in its ability to record the distribution of light corresponding to the scene. The image of an illuminated slit does not have the sharpness and definition of the source, and if t is now regarded as a spatial variable, a slightly blurred image is a consequence of the convolution operation. Knowing the impulse response of the system (or camera), the image information can be processed to restore some of the definition.

These are both examples of deconvolution and the reason for the terminology is evident from the input–output relation for a system. From Eq. (3.40) we have

$$x(t) = \int_0^t h(t-\tau) f(\tau)\, d\tau, \tag{3.41}$$

and if, say, $x(t)$ and $h(t)$ are known, the task is to determine the function $f(t)$ which appears under the integral sign. Equation (3.41) is then an *integral equation* for $f(t)$ and its solution is possible because of the convolution theorem. By application of a Laplace transform

$$X(s) = H(s)F(s)$$

and hence

$$f(t) = \mathscr{L}^{-1}\{F(s)\} = \mathscr{L}^{-1}\left\{\frac{X(s)}{H(s)}\right\}. \tag{3.42}$$

The process is similar if the impulse response is the unknown quantity.

Example. If the input $tu(t)$ applied to a system at rest produces the output $\frac{1}{8}(2t - \sin 2t)u(t)$, find the impulse response of the system.

Since

$$x(t) = \tfrac{1}{8}(2t - \sin 2t)u(t),$$

we have

$$X(s) = \frac{1}{4}\left(\frac{1}{s^2} - \frac{1}{s^2+4}\right) = \frac{1}{s^2(s^2+4)}$$

and

$$F(s)=\frac{1}{s^2}.$$

Hence

$$H(s)=\frac{X(s)}{F(s)}=\frac{1}{s^2+4},$$

giving

$$h(t)=\tfrac{1}{2}(\sin 2t)u(t).$$

Interconnected systems

Large systems are made up of smaller ones, and the essence of the systems approach is the synthesis and analysis of complicated systems using simpler ones whose properties are more easily established and understood. The small ones are subsystems of the whole, but are also systems in their own right. Thus, a digital computer is made up of arithmetic units, control circuits, memory devices, and so forth, and each can be broken down into smaller electrical subsystems. To design and construct a large system, it is a practical necessity to use small ones as building blocks.

There are two basic ways in which systems can be interconnected: parallel and series (or cascade). These are shown in Fig. 3.13. With the convention that at a pick-off point, such as the left-hand path division in Fig. 3.13(a), the signal proceeds unaltered along each path, whereas at a summing point where paths come together the signals are added, it is evident that for *two systems in parallel*:

$$\begin{aligned}x(t)&=h_1(t)*f(t)+h_2(t)*f(t)\\&=h(t)*f(t)\end{aligned}$$

with

$$h(t)=h_1(t)+h_2(t); \tag{3.43}$$

for *two systems in series*:

$$x(t)=h_2(t)*\{h_1(t)*f(t)\}.$$

Since successive convolutions obey the associative law,

$$x(t)=h(t)*f(t)$$

with

$$h(t)=h_1(t)*h_2(t). \tag{3.44}$$

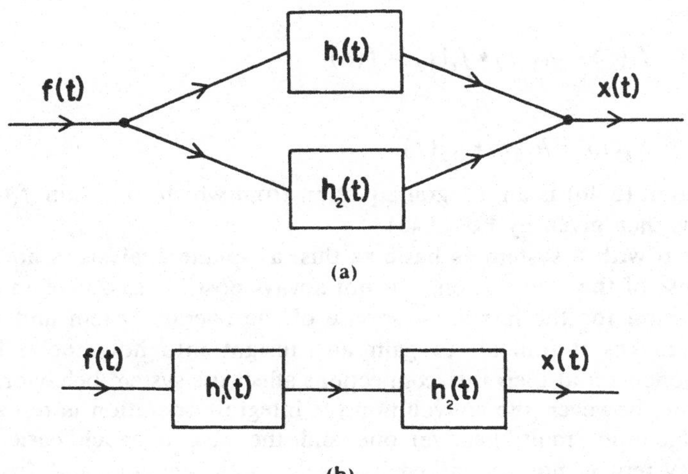

Fig. 3.13: (a) Parallel and (b) series connections.

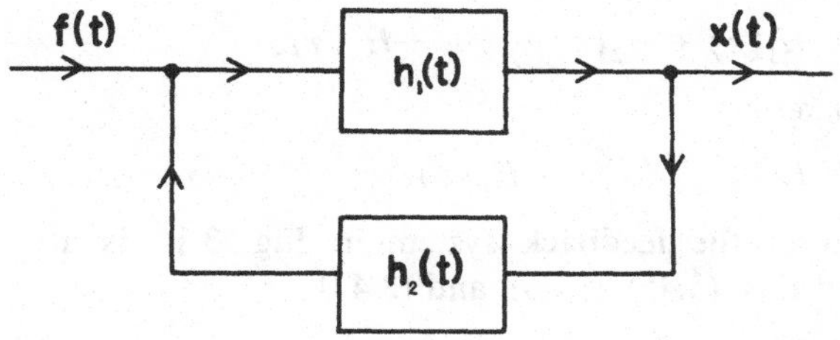

Fig. 3.14: A feedback system.

Whether joined in parallel or series, the two systems can be represented as a single one whose impulse response is given by Eq. (3.43) or (3.44), but as more systems are joined in series, the resulting expression becomes increasingly involved.

As another example, consider the feedback system shown in Fig. 3.14. If $f_1(t)$ is the input to the upper subsystem, the output from this is $h_1(t) * f_1(t)$, which is also the overall output, that is,

$$x(t) = h_1(t) * f_1(t), \tag{3.45}$$

as well as the input to the lower subsystem. The output from the latter is therefore $h_2(t) * \{h_1(t) * f_1(t)\}$, and at the summing point on the left we have

$$f(t) + h_2(t) * \{h_1(t) * f_1(t)\} = f_1(t);$$

that is,

$$f_1(t) - h_3(t) * f_1(t) = f(t) \tag{3.46}$$

with

$$h_3(t) = h_1(t) * h_2(t). \tag{3.47}$$

Equation (3.46) is an integral equation from which to obtain $f_1(t)$, and $x(t)$ is then given by Eq. (3.45).

Even with a system as basic as this, a t-plane analysis is not trivial. Because of the convolution, it is not always possible to derive an explicit expression for the impulse response of the overall system and this, in turn, makes it difficult to gain any insight into how the individual components and their interconnections affect the system behavior. In the s plane, however, the convolution (or integral) operation is replaced by an algebraic (multiplicative) one and the task is much easier. Each (sub)system is now described by its transfer function, and from Eqs. (3.43) and (3.44)

for *n systems in parallel*:

$$H(s) = H_1(s) + H_2(s) + \cdots + H_n(s); \tag{3.48}$$

for *n systems in series*:

$$H(s) = H_1(s) H_3(s) \cdots H_n(s). \tag{3.49}$$

The description of the feedback system in Fig. 3.14 is also greatly simplified. From Eqs. (3.45), (3.46), and (3.47)

$$X(s) = H_1(s) F_1(s)$$

and

$$F_1(s)\{1 - H_1(s) H_2(s)\} = F(s),$$

giving

$$X(s) = H(s) F(s),$$

where the overall transfer function $H(s)$ is

$$H(s) = \frac{H_1(s)}{1 - H_1(s) H_2(s)}. \tag{3.50}$$

In contrast to the situation in the t plane, the combination of (sub)systems into a single system is always possible and straightforward in the s plane, and the resulting *s-plane algebra* leads naturally to the idea of *block diagrams* in which a system is represented by a block having a designated transfer function. The blocks are interconnected with lines whose arrows show the direction of signal flow, and the manner in which signals are combined at a junction is indicated. A block may represent a single subsystem or component, or several, and blocks in a complicated system can be combined or manipulated to yield a simplified block diagram. Thus, the feedback system of Fig. 3.14 can be represented by a single block whose transfer function is given in Eq. (3.50). Conversely, for a complicated system the analytic form of the transfer function indicates the manner in which the system can be broken down into simpler and more easily constructed subsystems.

Exercises

*1. If the input $f(t) = 4t^3u(t)$ applied to a system at rest produces the output

$$x(t) = \left(t^3 - \tfrac{15}{2}t + 8\sin t - \tfrac{1}{4}\sin 2t\right)u(t),$$

find the impulse response of the system.

2. If the output of a system whose impulse response is $h(t) = e^{-t}(\sin t)u(t)$ is

$$x(t) = \tfrac{1}{5}(e^{-3t} - e^{-t}\cos t + 2e^{-t}\sin t)u(t),$$

find the input, assuming the system was initially at rest.

3. If the output of a system described by the differential equation $(D^2 + 1)x(t) = f(t)$ is

$$x(t) = (t - \sin t)u(t) + \{\cos(t-1) + \sin(t-1) - t\}u(t-1),$$

find the input $f(t)$, assuming the system was initially at rest.

4. Given that

$$f(t) + 3a\int_0^t \sin a(t-\tau)f(\tau)\,d\tau = (\cos at)u(t),$$

where a is a constant, find $f(t)$.

5. Solve the integral equation

$$f(t) + 5\int_0^t (e^{-\tau} - e^{-3\tau})f(t-\tau)\,d\tau = 9te^{-2t}u(t)$$

to obtain $f(t)$.

6. Solve the integral equation

$$3f(t) + \int_0^t (e^{\tau} - e^{-\tau} - \sqrt{2}\sin\tau\sqrt{2})f(t-\tau)\,d\tau = \tfrac{3}{2}(e^{t} - e^{-t})u(t)$$

to obtain $f(t)$. [*Hint*: Express $F(s)$ as a function of $s + 2$.]

3.8 Simultaneous differential equations

To analyze a relatively complicated system like a multisegment electrical circuit, it may be convenient to introduce additional unknowns such as the currents or voltages in individual segments which are then related to one another using the conditions obtaining at the junctions of the segments. The result is a set of m simultaneous equations in m unknowns, only one of which is the quantity originally desired, for example, the output $x(t)$ of the entire system. It often happens that the highest derivative in each equation is the first, but this is not essential, nor is it required that each equation involve every unknown.

By differentiating the equations appropriately, we can eliminate all of the unknowns except $x(t)$, leaving a single differential equation which can be solved using Laplace transforms. Alternatively, the Laplace-transform technique can be applied to the original set of coupled equations, and it is usually simpler to proceed in this manner.

Example. Solve

$$(D+6)x_1(t) + 5x_2(t) = 0,$$

$$3x_1(t) - (D-2)x_2(t) = -2e^t u(t)$$

for $x_1(t)$ and $x_2(t)$, with $x_1(0) = 0$, $x_2(0) = 1$.

Assuming $x_1(t)$ and $x_2(t)$ are both zero for $t < 0$, application of a Laplace transform gives

$$(s+6)X_1(s) + 5X_2(s) = 0,$$

$$3X_1(s) - (s-2)X_2(s) = -1 - \frac{2}{s-1} = -\frac{s+1}{s-1}.$$

These are two algebraic equations for $X_1(s)$ and $X_2(s)$. $X_2(s)$ can be eliminated by multiplying the first equation by $s-2$, the second by 5, and adding. Then

$$\{(s-2)(s+6)+15\}X_1(s) = -5\frac{s+1}{s-1},$$

from which we have

$$X_1(s) = -\frac{5}{(s-1)(s+3)} = -\frac{5}{4}\left(\frac{1}{s-1} - \frac{1}{s+3}\right).$$

Hence

$$x_1(t) = -\tfrac{5}{4}(e^t - e^{-3t})u(t).$$

To obtain $X_2(s)$ we can return to the transformed equations and eliminate $X_1(s)$, or simply note that the first equation implies

$$X_2(s) = -\frac{s+6}{5}X_1(s)$$

$$= \frac{s+6}{(s-1)(s+3)} = \frac{1}{4}\left(\frac{7}{s-1} - \frac{3}{s+3}\right).$$

Thus

$$x_2(t) = \tfrac{1}{4}(7e^t - 3e^{-3t})u(t),$$

which completes the solution.

If, instead, we had eliminated $x_2(t)$ from the original differential equations, the equation that we would have obtained for $x_1(t)$ is

$$(D^2 + 4D + 3)x_1(t) = -10e^t u(t)$$

with $x_1(0) = 0$ and $x_1'(0) = -5$, and the corresponding equation for $x_2(t)$ is

$$(D^2 + 4D + 3)x_2(t) = 14e^t u(t) + 2\delta(t)$$

with $x_2(0) = 1$ and $x_2'(0) = 4$.

This simple example illustrates several features of any set of simultaneous differential equations. The characteristic functions of the higher-order equations satisfied by the individual unknowns are the same and equal to the characteristic function $V(s)$ of the overall system. $V(s)$ is proportional to the determinant of the coefficients in the transformed set of coupled equations. The highest power of s in the expression for $V(s)$ is, of course, the order of the system, and this is less than or equal to the sum of the highest derivatives in each of the m simultaneous differential equations.

It is obvious that we can also break up an nth-order differential equation into a set of m $(\leq n)$ simultaneous differential equations by introducing $m - 1$ additional unknowns which may or may not have any physical significance themselves. The decomposition is not unique, but all sets uniquely determine the original unknown $x(t)$. If such a decomposition is carried out, there are advantages to having $m = n$ with each differential equation of the first order. In particular, for a first-order equation the determination of the equivalent input is a trivial task. As an example, if

$$(D + 2)x(t) = f(t)$$

with $x(0) = a$, the equivalent input is simply

$$\tilde{f}(t) = f(t) + a\delta(t).$$

For a system which is effectively at rest, the techniques discussed in the previous two sections are available.

Vector formulation

Any nth-order differential equation or set of simultaneous equations representing an nth-order system can be expressed as a set of n first-order equations in n unknowns, one of which could be the actual output of the system, but does not have to be. By appropriate manipulation, the equations can be written so that each involves the derivative of just a single unknown. The form of the mth equation is then

$$Dx_m(t) = A_{m1}x_1(t) + A_{m2}x_2(t) + \cdots + A_{mn}x_n(t) + \tilde{f}_m(t),$$

where the A_{mi} are constants and $\tilde{f}_m(t)$ is the equivalent input, incorporating the effect of a nonzero initial value of $x_m(t)$. The entire set expressed in vector notation is

$$\overrightarrow{Dx(t)} = \bar{\bar{A}} \cdot \overrightarrow{x(t)} + \overrightarrow{\tilde{f}(t)}, \qquad (3.51)$$

where $\overrightarrow{Dx(t)}$, $\overrightarrow{x(t)}$, and $\overrightarrow{\tilde{f}(t)}$ are n-dimensional vectors and $\bar{\bar{A}}$ is an $n \times n$ matrix with constant elements.

Some of the advantages of this standardized form are:

(i) It is completely general. Any differential equation or set of simultaneous equations can be written in this way, and the order of the system affects only the dimensions of the vectors and matrix.

(ii) It is a trivial matter to incorporate the effect of nonzero initial values.

(iii) Existing computer codes for the numerical solution of differential equations are based on this form.

Example. If

$$3Dx_1(t) + (2D - 1)x_2(t) = e^t u(t),$$
$$(D + 1)x_1(t) + (D - 1)x_2(t) = e^{-t}u(t)$$

with $x_1(0) = 2$ and $x_2(0) = 1$, express the equations in vector form.

As the equations stand, each involves more than one derivative. However, by subtracting twice the second from the first,

$$(D - 2)x_1(t) + x_2(t) = (e^t - 2e^{-t})u(t),$$

and by subtracting the first from three times the second,

$$3x_1(t) + (D-2)x_2(t) = (3e^{-t} - e^{t})u(t).$$

Hence

$$Dx_1(t) = 2x_1(t) - x_2(t) + (e^{t} - 2e^{-t})u(t),$$
$$Dx_2(t) = -3x_1(t) + 2x_2(t) + (3e^{-t} - e^{t})u(t),$$

that is,

$$\overrightarrow{Dx(t)} = \bar{\bar{A}} \cdot \overrightarrow{x(t)} + \overrightarrow{\tilde{f}(t)},$$

where

$$\overrightarrow{x(t)} = \{x_1(t), x_2(t)\},$$
$$\bar{\bar{A}} = \begin{pmatrix} 2 & -1 \\ -3 & 2 \end{pmatrix},$$

and

$$\overrightarrow{\tilde{f}(t)} = \{(e^{t} - 2e^{-t})u(t) + 2\delta(t), (3e^{-t} - e^{t})u(t) + \delta(t)\}.$$

Note that the insertion of the initial values was deferred until the last stage of the analysis.

In this example the desired choice of the two unknowns describing the second-order system was evident from the original form of the differential equations. For a general nth-order system specified either by a single differential equation or in some other manner, there is considerable freedom in how the unknowns are chosen, and the essence of the *state-variable* approach to systems is the establishment of criteria for their selection.

System decomposition

When analyzing a complicated electrical circuit, it is often convenient to introduce unknowns (or variables) additional to the one being sought, and this practice is customary in many system problems. In some instances these new unknowns are themselves outputs from subsystems making up the whole, and constitute intermediate outputs which are of interest in their own right. But even if they are of no real interest themselves, it may still be advantageous to introduce them.

As an example, consider the system whose transfer function is

$$H(s) = \frac{1}{s(s^2+1)^2}.$$

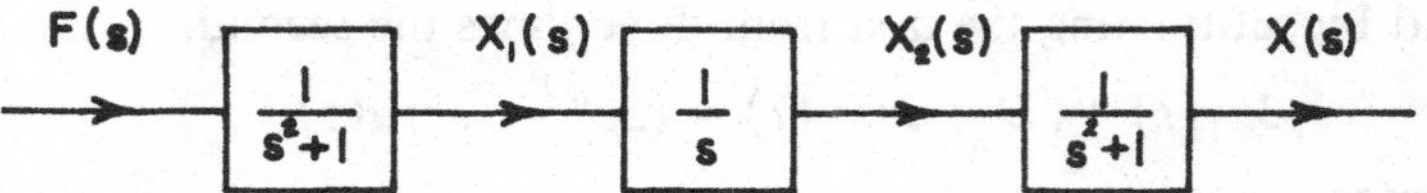

Fig. 3.15: Possible decomposition of a system with transfer function $1/[s(s^2+1)^2]$.

An equivalent representation of the system is shown in Fig. 3.15, and if $x_1(t)$ and $x_2(t)$ are the intermediate outputs from the first and second subsystems, respectively,

$$X_1(s) = \frac{1}{s^2+1}F(s), \qquad X_2(s) = \frac{1}{s}X_1(s), \qquad X(s) = \frac{1}{s^2+1}X_2(s).$$

In the particular case when $f(t) = \delta(t)$ so that $F(s) = 1$,

$$x_1(t) = \mathscr{L}^{-1}\left\{\frac{1}{s^2+1}\right\} = (\sin t)u(t)$$

and

$$x_2(t) = D_0^{-1}x_1(t)$$
$$= \int_0^t \sin t'\,dt'\,u(t) = (1 - \cos t)u(t).$$

The output of the overall system is then

$$x(t) = \{(\sin t)u(t)\} * \{(1-\cos t)u(t)\}$$
$$= \int_0^t (1-\cos\tau)\sin(t-\tau)\,d\tau\,u(t),$$

which can be evaluated by expanding $\sin(t-\tau)$ and using the double-angle formulas for the trigonometric functions. The result is

$$x(t) = \left(1 - \cos t - \tfrac{1}{2}t\sin t\right)u(t),$$

and in effect the decomposition of the system has enabled us to show

$$\mathscr{L}^{-1}\left\{\frac{1}{s(s^2+1)^2}\right\} = \left(1 - \cos t - \frac{1}{2}t\sin t\right)u(t).$$

When systems are combined to constitute a larger and more complex system, it may happen that the resulting transfer function $H(s)$ is *not* the reciprocal of a polynomial in s, implying that the overall system cannot be described by a differential equation. The feedback system in Fig. 3.14 is a case in point, and even if each of $H_1(s)$ and $H_2(s)$ has the form $1/[V(s)]$ where $V(s)$ is a polynomial, the transfer function $H(s)$ in Eq.

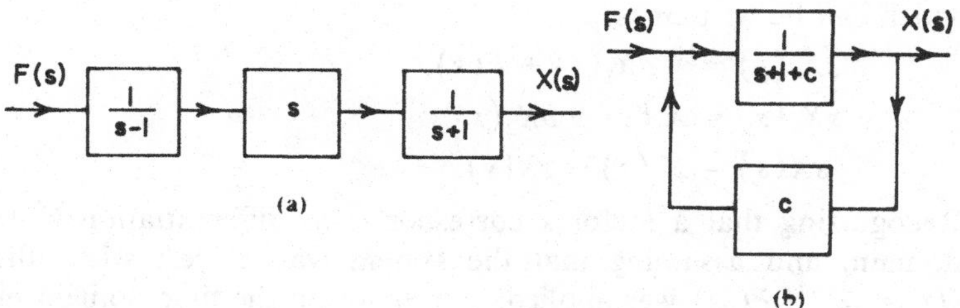

Fig. 3.16: Possible decompositions of systems with transfer functions (a) $s/(s^2 - 1)$ and (b) $1/(s + 1)$.

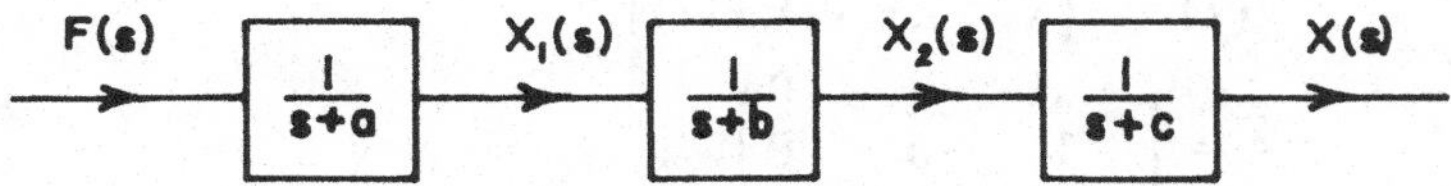

Fig. 3.17: Possible decomposition of a system with transfer function $1/[(s + a)(s + b)(s + c)]$.

(3.50) will not in general have this form. For example, the system with $H(s) = s/(s^2 - 1)$ can be broken up as shown in Fig. 3.16(a), and the middle (sub)system then acts as a differentiator. We could even represent the right-hand subsystem whose transfer function is $1/(s + 1)$ as the feedback system in Fig. 3.16(b) with any value of c.

As previously noted, an nth-order differential equation can be broken up into n simultaneous differential equations by introducing $n - 1$ additional unknowns, and the resulting set of equations can then be expressed in vector form. This implies the decomposition of a system into subsystems whose transfer functions involve s to the first power at most. As an example, consider the system with transfer function

$$H(s) = \frac{1}{(s + a)(s + b)(s + c)}.$$

The differential equation describing this system is clearly third order, but a possible decomposition of the system is shown in Fig. 3.17.

If $X_1(s)$ and $X_2(s)$ are the outputs of the first and second subsystems, respectively, we have

$$X_1(s) = \frac{1}{s + a}F(s), \qquad X_2(s) = \frac{1}{s + b}X_1(s), \qquad X(s) = \frac{1}{s + c}X_2(s),$$

which can be written as

$$sX_1(s) = -aX_1(s) + F(s),$$
$$sX_2(s) = X_1(s) - bX_2(s),$$
$$sX(s) = X_2(s) - cX(s).$$

Recognizing that a factor s corresponds to differentiation in the time domain, and assuming that the system was at rest when the input $f(t) = \mathscr{L}^{-1}\{F(s)\}$ was applied, conversion to the time domain produces a set of equations that are already in vector form, namely,

$$\overrightarrow{Dx(t)} = \overline{\overline{A}} \cdot \overrightarrow{x(t)} + \overrightarrow{f(t)},$$

where

$$\overrightarrow{x(t)} = \{x_1(t), x_2(t), x(t)\},$$
$$\overline{\overline{A}} = \begin{pmatrix} -a & 0 & 0 \\ 1 & -b & 0 \\ 0 & 1 & -c \end{pmatrix},$$

and

$$\overrightarrow{f(t)} = \{f(t), 0, 0\}.$$

The three unknowns $x_1(t)$, $x_2(t)$, and $x(t)$ constitute a set of state variables defined as a minimal set of unknowns (or variables) sufficient to determine the response of the system to any given input. In this example, all three variables are outputs, but it is not necessary for a set of state variables to have this interpretation.

Exercises

Solve the following simultaneous differential equations for $x_1(t)$ and $x_2(t)$.

*1. $(D-2)x_1(t) + 3x_2(t) = 0,$

$2x_1(t) + (D-1)x_2(t) = 0$

with $x_1(0) = 8$, $x_2(0) = 3$.

2. $(2D+4)x_1(t) + (D-1)x_2(t) = e^{-3t}u(t),$

$(D+2)x_1(t) + Dx_2(t) = 0$

with $x_1(0) = 5$, $x_2(0) = 2$.

3. $x_1(t) - Dx_2(t) = 2e^{-t}u(t),$

$Dx_1(t) + x_2(t) = 4e^{t}u(t)$

with $x_1(0) = 2$, $x_2(0) = 0$.

4. $(2D - 1)x_1(t) + (D + 13)x_2(t) = 96e^{2t}u(t),$

$(2D - 3)x_1(t) - (D - 7)x_2(t) = 0$

with $x_1(0) = 1,\ x_2(0) = -1.$

5. $(D^2 + 3)x_1(t) + Dx_2(t) = 15e^{-t}u(t),$

$4Dx_1(t) - (D^2 + 3)x_2(t) = -15(\sin 2t)u(t)$

with $x_1(0) = 35,\ x_1'(0) = -48,\ x_2(0) = 27,\ x_2'(0) = -55.$

Express the following equations in standard vector form using the same unknowns.

6. $3Dx_1(t) + (2D + 1)x_2(t) = 2e^{-t}u(t),$

$(D + 1)x_1(t) + Dx_2(t) = u(t)$

with $x_1(0) = 2,\ x_2(0) = 3.$

3.9 Stability

An important property of any system is its stability. Consider, for example, the feedback system shown in Fig. 3.14. If this is intended as an output-control device, the designer must balance the conflicting needs of sensitivity and stability. By increasing the amplification of the error signal in the feedback loop, the accuracy of control is initially improved, but too much amplification can lead to an overshoot that, if large enough, can cause the output to oscillate with increasing amplitude. With an actual device, internal losses will ensure that the output remains finite, but an unstable system of this type cannot perform a useful control function.

The stability of any system is determined by the possible behavior of the output $x(t)$ as $t \to \infty$. It is an attribute of the system itself and is independent of the input and the initial values (abbreviated iv), assumed finite. Three cases are distinguishable:

(i) If for one or more iv

$$\lim_{t\to\infty} |x(t)| = \infty,$$

the system is *unstable*.

(ii) If for all iv the limit is finite, but for one or more iv

$$\lim_{t\to\infty} |x(t)| \neq 0,$$

the system is *conditionally stable*.

(iii) If for all iv

$$\lim_{t\to\infty} |x(t)| = 0,$$

the system is *stable*.

No stable system can resonate, and for practical purposes a conditionally stable system is no more usable than an unstable one. The slightest variation in its parameters could cause the system to become unstable, and for this reason a conditionally stable system is sometimes referred to as conditionally *un*stable.

The criteria listed previously can be expressed in terms of the zeros of the characteristic function $V(s)$ describing the system. To see this, consider the homogeneous differential equation

$$\{D^2 - (s_1 + s_2)D + s_1 s_2\}x(t) = 0,$$

where s_1 and s_2 are constants. We seek the solution $x(t)$ whose initial values are $x(0) = a$, $x'(0) = b$, with a, b finite. Assuming $x(t) = 0$ for $t < 0$, application of a Laplace transform gives

$$X(s) = H(s)A(s),$$

where

$$A(s) = as + b - (s_1 + s_2)a$$

is the first-order polynomial contributed by the initial values, and

$$H(s) = \frac{1}{V(s)} = \frac{1}{(s - s_1)(s - s_2)}.$$

The constants s_1 and s_2 are therefore the characteristic roots. If $s_2 \neq s_1$

$$X(s) = \frac{C_1}{s - s_1} + \frac{C_2}{s - s_2},$$

where

$$C_1 = \frac{A(s_1)}{s_1 - s_2} \quad \text{and} \quad C_2 = -\frac{A(s_2)}{s_1 - s_2},$$

from which we have

$$x(t) = \left(C_1 e^{s_1 t} + C_2 e^{s_2 t}\right)u(t); \tag{3.52}$$

but if $s_2 = s_1$,

$$X(s) = \frac{C_1}{s - s_1} + \frac{C_2}{(s - s_1)^2}$$

with

$$C_1 = A'(s_1) \quad \text{and} \quad C_2 = A(s_1),$$

giving

$$x(t) = (C_1 + tC_2)e^{s_1 t}u(t). \tag{3.53}$$

From Eqs. (3.52) and (3.53) it is clear that if $\operatorname{Re} s_1$ and/or $\operatorname{Re} s_2 > 0$ the system is unstable, whereas if $\operatorname{Re} s_1$ and $\operatorname{Re} s_2 < 0$ the system is stable. If $\operatorname{Re} s_1 = 0$ and $\operatorname{Re} s_2 < 0$ (or vice versa), or $\operatorname{Re} s_1$ and $\operatorname{Re} s_2 = 0$ with $s_1 \neq s_2$, Eq. (3.52) shows that the system is conditionally stable, but if $s_1 = s_2$ with $\operatorname{Re} s_1 = 0$ the system is unstable because of the factor t in the solution of Eq. (3.53). Since s_1 and s_2 are also the singularities of the transfer function $H(s)$, these results can be expressed in terms of their locations in the complex s plane and extended to any system. The criteria for stability are then:

Unstable: one or more singularities in the right-half plane ($\operatorname{Re} s > 0$) or a nonsimple (i.e., repeated) singularity on the imaginary axis.
Conditionally stable: one or more simple singularities on the imaginary axis and the rest in the left-half plane ($\operatorname{Re} s < 0$).
Stable: all singularities in the left-half plane.

The application of these criteria is illustrated by the following examples.

If

$$H(s) = \frac{1}{s^2 + 1},$$

the singularities are $s = \pm i$: conditionally stable.

If

$$H(s) = \frac{1}{(s^2 + 1)^2},$$

the singularities are $s = \pm i$ (repeated): unstable (because of the nonsimple singularities).

If

$$H(s) = \frac{1}{s^4 - 1},$$

the singularities are $s = \pm 1, \pm i$: unstable (because of the singularity $s = 1$).

If

$$H(s) = \frac{1}{(s + 1)^2 + 1},$$

the singularities are $s = -1 \pm i$: stable.

The determination of the singularities or, equivalently, the characteristic roots, requires the solution of an algebraic equation. This can be a very time-consuming task in any practical situation. However, it is not necessary to obtain these precisely if only the stability is being investigated. It is sufficient to find the signs of the real parts of the roots, and there are methods available to provide this information.

3.10 Steady-state response

The only systems of practical concern are stable ones, and we now seek the response of a stable system to a particular periodic input. The input considered is

$$f(t) = e^{i\omega t}u(t) \tag{3.54}$$

with ω real. Since

$$e^{i\omega t} = \cos \omega t + i \sin \omega t,$$

$e^{i\omega t}$ is merely the complex phasor representation of the more familiar periodic functions $\cos \omega t$ and $\sin \omega t$.

To keep the discussion as simple as possible, the transfer function of the system is taken to be

$$H(s) = \frac{1}{(s - s_1)(s - s_2)}$$

as used in the previous section. It is assumed that $s_2 \neq s_1$ and $\operatorname{Re} s_1, \operatorname{Re} s_2 < 0$ in accordance with the requirements for a stable system. The Laplace transform of the output is then

$$X(s) = H(s)\{A(s) + F(s)\},$$

where $A(s)$ is a first-order polynomial in s produced by the initial values and

$$F(s) = \frac{1}{s - i\omega}.$$

By expansion in partial fractions, we obtain

$$H(s)A(s) = \frac{C_1}{s - s_1} + \frac{C_2}{s - s_2}$$

with

$$C_1 = \frac{A(s_1)}{s_1 - s_2}, \qquad C_2 = -\frac{A(s_2)}{s_1 - s_2},$$

and

$$H(s)F(s) = \frac{C_3}{s - s_1} + \frac{C_4}{s - s_2} + \frac{H(i\omega)}{s - i\omega}$$

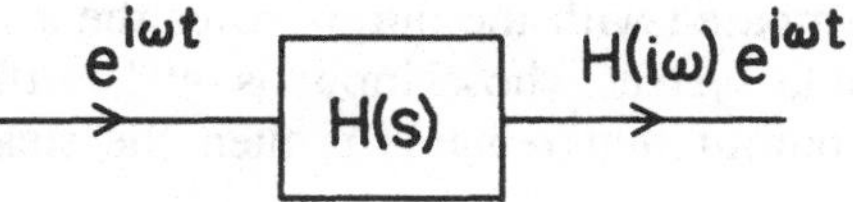

Fig. 3.18: The steady-state response.

with

$$C_3 = \frac{1}{(s_1 - i\omega)(s_1 - s_2)}, \qquad C_4 = -\frac{1}{(s_2 - i\omega)(s_1 - s_2)}.$$

The output is therefore

$$x(t) = \left\{C_1 e^{s_1 t} + C_2 e^{s_2 t} + C_3 e^{s_1 t} + C_4 e^{s_2 t} + H(i\omega)e^{i\omega t}\right\} u(t).$$

The first two terms in this expression are caused by the nonzero initial values and constitute an exponentially decaying output resulting from the situation that obtained for $t < 0$. The next two terms also decay with increasing t and are produced by the abrupt turn-on of the input at $t = 0$. Thus, for a stable system, all four terms represent *transients* which die out as t increases. Their rates of decay depend on the magnitudes of the negative real parts of s_1 and s_2, and these are, in turn, fixed by the parameters of the system. The rates are independent of the input and the initial values, but the amplitudes of the transients do depend on these quantities.

The fifth and final term in the output has the same mathematical form as the input $f(t)$, and since $H(i\omega)$ is merely a complex constant, this term does not decay with time. It is the steady-state term corresponding to a particular integral of the differential equation describing the system, and constitutes the *steady-state response*, namely,

$$x(t) = H(i\omega)e^{i\omega t} \tag{3.55}$$

(the inclusion of the unit-step function is unnecessary since the output is only meaningful for $t \gg 0$). In contrast to the transients or to the complete output, which can be determined only by the evaluation of a convolution integral or, in practice, by using the Laplace-transform technique, the steady-state response is obtained immediately from a knowledge of the transfer function of the system. No inverse transformation is involved despite the fact that the transfer function is an s-plane characteristic of the system. The steady-state response depends on the transfer function and the frequency ω of the periodic exponential input, and nothing else. This is illustrated in Fig. 3.18.

For many applications the steady-state response is the only output of interest. The relatively brief effects of turning on the input (or the system itself) are of no concern compared with the output after the system has "settled down." For a stable system whose input is $e^{i\omega t}$, settling will eventually occur, and the output that remains is then the steady-state response given in Eq. (3.55).

Example. If the input $\cos\omega t$ is applied to a system whose transfer function is $H(s) = 1/(s+1)$, find the steady-state response.

Since

$$\cos\omega t = \tfrac{1}{2}(e^{i\omega t} + e^{-i\omega t}),$$

the steady-state response is, by superposition,

$$\begin{aligned} x(t) &= \frac{1}{2}\{H(i\omega)e^{i\omega t} + H(-i\omega)e^{-i\omega t}\} \\ &= \frac{1}{2}\left\{\frac{e^{i\omega t}}{1+i\omega} + \frac{e^{-i\omega t}}{1-i\omega}\right\} \\ &= \frac{1}{1+\omega^2}(\cos\omega t + \omega\sin\omega t), \end{aligned}$$

which is real since the input is real. In particular, if $f(t) = 1$, the steady-state response can be obtained by putting $\omega = 0$ and is $x(t) = 1$. Similarly, if

$$f(t) = \sin\omega t = \frac{1}{2i}(e^{i\omega t} - e^{-i\omega t}),$$

the response is

$$x(t) = \frac{1}{1+\omega^2}(\sin\omega t - \omega\cos\omega t).$$

More generally, if the periodic input

$$\begin{aligned} f(t) &= \sum_{n=1}^{N}\{a_n\cos n\omega t + b_n\sin n\omega t\} \\ &= \sum_{n=1}^{N}\{\tfrac{1}{2}(a_n - ib_n)e^{in\omega t} + \tfrac{1}{2}(a_n + ib_n)e^{-in\omega t}\} \end{aligned}$$

is applied to a system whose transfer function is $H(s)$, the steady-state response is

$$x(t) = \sum_{n=1}^{N}\{\tfrac{1}{2}(a_n - ib_n)H(in\omega)e^{in\omega t} + \tfrac{1}{2}(a_n + ib_n)H(-in\omega)e^{-in\omega t}\}.$$

Exercises

1. Determine the stability of the systems described by the differential equations
   *(i) $(D^2 + 2D + 1)x(t) = f(t)$.
   (ii) $(D^3 + 1)x(t) = f(t)$.
   (iii) $(D^3 + D^2 + D + 1)x(t) = f(t)$.
   (iv) $(D + 1 + D_0^{-1})x(t) = f(t)$.

2. Determine the stability of the systems described by the differential equations
   (i) $(2D - 1)x_1(t) + (D + 13)x_2(t) = f_1(t)$,
   $(2D - 3)x_1(t) - (D - 7)x_2(t) = f_2(t)$.
   (ii) $(D + 2)x_1(t) + (D - 2)x_2(t) = f_1(t)$,
   $(D - 1)x_1(t) + (2D + 3)x_2(t) = f_2(t)$.

3. If $V(s) = s^3 + (1 + 2a)(s^2 + s) + 1$, determine the stability of the system as a function of the real parameter a, $-\infty < a < \infty$.

4. Given the equation

$$f(t) = x(t) + a\int_0^t (1 - e^{-2\tau})x(t - \tau)\, d\tau:$$

   *(i) Find $x(t)$ if $f(t) = tu(t)$ and $a = \frac{1}{2}$.
   (ii) Determine the stability of the system whose input is $f(t)$ and output is $x(t)$ as a function of the real parameter a, $0 \leq a < \infty$.

5. Prove that the system described by the differential equation

$$\left(D^2 + a_1 D + a_2\right)x(t) = f(t),$$

   where a_1, a_2 are real, is stable if and only if $a_1 > 0$, $a_2 > 0$.

6. If the input $e^{i\omega t}$ is applied to a system whose transfer function is

$$H(s) = \frac{1}{(s + 1)^2 + 1}:$$

   (i) Find the steady-state response.
   (ii) Deduce the responses for the inputs $\cos \omega t$ and $\sin \omega t$, and express these in real form.

Suggested reading

Kaplan, W. *Operational Methods for Linear Systems*. Addison-Wesley, Reading, MA, 1962. A more rigorous presentation of the mathematical methods that are the basis of linear-systems theory.

Cheng, D. K. *Analysis of Linear Systems*. Addison-Wesley, Reading, MA, 1959. A good coverage with emphasis on the practical aspects.

Liu, C. L. and J. W. S. Liu. *Linear Systems Analysis*. McGraw-Hill, New York, 1975. A more advanced operational presentation which includes a discussion of the state-space (or variables) description of systems.

Lighthill, M. J. *An Introduction to Fourier Analysis and Generalized Functions*. Cambridge University Press, Cambridge, UK, 1958. A condensed but rigorous derivation of the mathematical properties of the impulse function.

CHAPTER 4

Fourier series

The 1822 publication by the French mathematician Joseph Fourier (1768–1830) of his *Analytical Theory of Heat* is a milestone in the history of pure and applied mathematics. The work documented the claim made earlier (in 1807) by Fourier that an arbitrary function $f(t)$ defined in $(-\pi, \pi)$ can be represented by a trigonometric series of the form

$$\sum_{n=0}^{\infty} (a_n \cos nt + b_n \sin nt),$$

with coefficients that can be obtained from a knowledge of $f(t)$. Since $f(t)$ can even be discontinuous, the idea that it could be represented using cosines and sines, which are entirely smooth functions, would seem to defy logic, and met with considerable opposition at the time. As a matter of fact, it is not completely true and, as others showed later, mild restrictions must be placed on $f(t)$ to ensure convergence of the series. Nevertheless, Fourier's discovery had far-reaching consequences in most branches of science. It is the source of the concepts that underlie the language and thinking of all electrical engineers, and the representation of signals as combinations of trigonometric functions is fundamental to the discipline.

William Thompson (later Lord Kelvin) said that on 1 May 1840 (when he was only sixteen) "I took Fourier out of the University Library; and in a fortnight I had mastered it – gone right through it." The reader should aim to do no less.

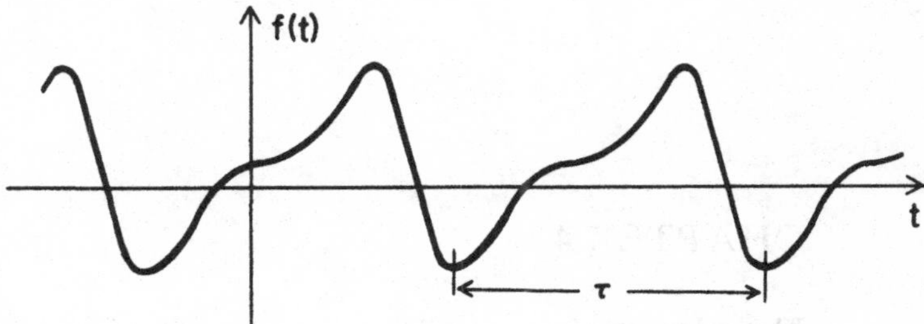

Fig. 4.1: Periodic function.

4.1 Periodic functions

A function $f(t)$ defined for all t is *periodic* if there exists a positive number τ such that

$$f(t+\tau)=f(t) \tag{4.1}$$

for all t, and τ is then called a period of $f(t)$. Figure 4.1 is an example of such a function, and we observe that the graph can be obtained by periodic repetition of the information contained in any interval of length τ. From Eq. (4.1) it follows that if n is an integer,

$$f(t+n\tau)=f(t)$$

so that $2\tau, 3\tau, \ldots$ are also periods of $f(t)$. The smallest value of τ for which the equation is satisfied is called the *fundamental period* and is indicated in Fig. 4.1. According to the definition (4.1), a constant is a periodic function, but it does not have a fundamental period.

The simplest examples of periodic functions are sines and cosines. Since

$$\sin\omega\left(t+\frac{2\pi}{\omega}\right)=\sin\omega t\cos 2\pi+\cos\omega t\sin 2\pi=\sin\omega t,$$

$\sin\omega t$ is periodic with period $\tau=2\pi/\omega$, and this is, indeed, the fundamental period. Similarly, $\cos\omega t$ is periodic with the same period, and because

$$\cos\omega t=\sin\omega\left(t+\frac{\pi}{2\omega}\right)=\sin\omega\left(t+\frac{\tau}{4}\right),$$

the cosine differs from the sine only in a displacement of the origin of time by a quarter of the fundamental period. Both functions represent a single frequency or simple harmonic oscillation of unit amplitude, and could be produced by attaching a pen to the end of an arm rotating at ω

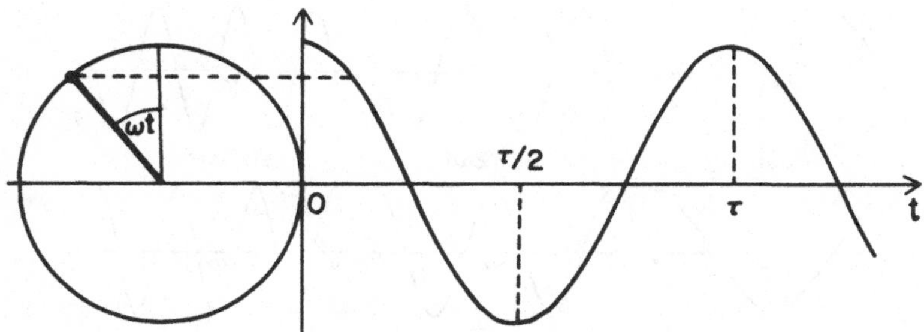

Fig. 4.2: Simple harmonic oscillation.

radians per second and moving the recording paper along (see Fig. 4.2). The time $2\pi/\omega$ required for one complete revolution ($= 2\pi$ radians) is likewise the period τ, and the angular frequency ω of $\sin \omega t$ and $\cos \omega t$ is therefore

$$\omega = \frac{2\pi}{\tau} \text{ rad/s.}$$

The (linear) frequency ν of a periodic function is the number of cycles or periods per second (i.e., $\nu = 1/\tau$), and the international unit for cycles per second is hertz, abbreviated Hz. Thus

$$\omega = 2\pi\nu, \tag{4.2}$$

and for the standard ac power supply in the United States $\nu = 60$ Hz, implying $\omega = 120\pi$ rad/s.

If n is an integer, the functions $\sin n\omega t$ and $\cos n\omega t$ also have period $\tau = 2\pi/\omega$, but this is not the fundamental period unless $n = 1$. The fundamental period is τ/n, and both functions represent a single frequency oscillation whose frequency is n times that of $\sin \omega t$ and $\cos \omega t$ (see Fig. 4.3). Frequencies that are integer multiples of one another are said to be harmonically related, and if $n > 1$, $\sin n\omega t$ and $\cos n\omega t$ constitute harmonics of the fundamental oscillations $\sin \omega t$ and $\cos \omega t$.

Periodic functions occur widely in nature and in all branches of science. The motion of the earth around the sun is periodic to a high degree of accuracy. The output of the human heart is periodic under normal conditions, and the signal produced at any point in space by a scanning radar is likewise periodic. Many periodic functions are quite complicated; therefore, it is desirable to represent them in terms of the simpler sines and cosines described above. To see how this could be

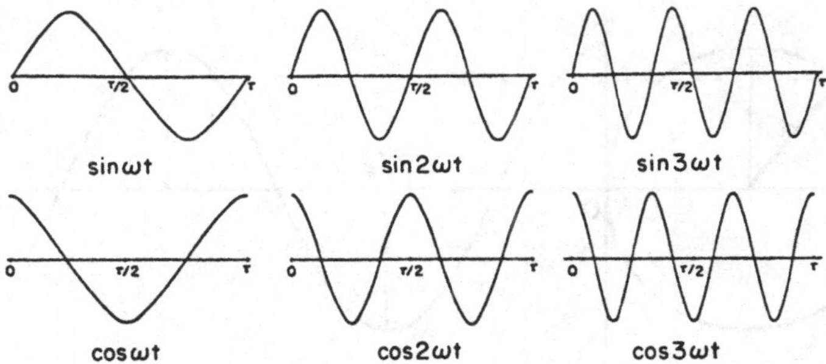

Fig. 4.3: Harmonically related functions.

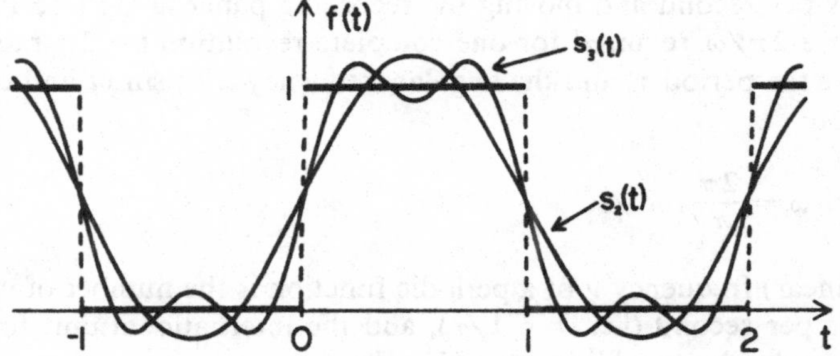

Fig. 4.4: Series approximations to a square wave.

done, consider the function $f(t)$ defined over a period as

$$f(t) = \begin{cases} 1 & \text{for } 0 < t < 1, \\ 0 & \text{for } 1 < t < 2. \end{cases}$$

This, together with its periodic extension, defines $f(t)$ for (almost) all t and the resulting square wave is shown in Fig. 4.4. The average height is 0.5, and if we take this as the first term of the representation, the difference is a waveform that vaguely resembles $\sin \pi t$. By inspection the appropriate amplitude is about 0.6, and a two-term approximation is then

$$s_2(t) = 0.5 + 0.6 \sin \pi t.$$

As seen from Fig. 4.4, this is beginning to take on the required shape, but it needs to be made more square. The agreement is improved by adding a third term $0.2 \sin 3\pi t$, leading to the three-term approximation

$$s_3(t) = 0.5 + 0.6 \sin \pi t + 0.2 \sin 3\pi t,$$

and by proceeding in this manner using the higher harmonics $\sin 5\pi t$, $\sin 7\pi t, \ldots,$ it seems possible that the square wave could be reproduced.

Examples of (finite) trigonometric series are $s_2(t)$ and $s_3(t)$. We shall now examine this type of series and develop formulas for the coefficients.

4.2 Trigonometric series

Consider a function $f(t)$ which can be written as

$$f(t) = \frac{1}{2}a_0 + \sum_{n=1}^{N} (a_n \cos n\omega t + b_n \sin n\omega t) \tag{4.3}$$

for all t where the coefficients a_0 and a_n, b_n are constants. If a_1 and b_1 are not both zero, the smallest period common to all of the terms on the right-hand side is $\tau = 2\pi/\omega$, which is the fundamental period of the terms corresponding to $n = 1$. Therefore $f(t)$ is a periodic function with period τ.

The coefficients in an expansion of this form can be expressed as weighted integrals of the function $f(t)$, a result which is made possible by the orthogonality of the trigonometric functions.

Orthogonality

The following identities are important and can be proved using the exponential forms for the sine and cosine functions:

$$\cos A \cos B = \tfrac{1}{2}\{\cos(A - B) + \cos(A + B)\},$$

$$\sin A \sin B = \tfrac{1}{2}\{\cos(A - B) - \cos(A + B)\},$$

$$\sin A \cos B = \tfrac{1}{2}\{\sin(A - B) + \sin(A + B)\}. \tag{4.4}$$

The basis functions for the expansion (4.3) are

$$1\ (= \cos 0\omega t) \quad \text{and} \quad \cos n\omega t, \sin n\omega t \qquad \text{with } 1 \le n \le N.$$

To show the effect of taking the product of any two members of the base and integrating over a time interval of length $\tau = 2\pi/\omega$, consider

$$\int_0^\tau \cos m\omega t \cos n\omega t \, dt,$$

where m and n are nonnegative integers. If $m \neq n$,

$$\begin{aligned}\int_0^{\tau} \cos m\omega t \cos n\omega t \, dt &= \int_0^{2\pi/\omega} \cos m\omega t \cos n\omega t \, dt \\ &= \frac{1}{\omega}\int_0^{2\pi} \cos mt' \cos nt' \, dt' \\ &= \frac{1}{2\omega}\int_0^{2\pi} \{\cos(m-n)t' + \cos(m+n)t'\} \, dt' \\ &= \frac{1}{2\omega}\left[\frac{\sin(m-n)t'}{(m-n)} + \frac{\sin(m+n)t'}{(m+n)}\right]_0^{2\pi} \\ &= 0,\end{aligned}$$

since the sine of any integer multiple of π is zero. If $m = n \neq 0$, the above result is incorrect because of the division by $m - n$ $(= 0)$ in the next to the last line, and in this case

$$\begin{aligned}\int_0^{\tau} \cos n\omega t \cos n\omega t \, dt &= \frac{1}{2\omega}\int_0^{2\pi} \{1 + \cos 2nt'\} \, dt' \\ &= \frac{1}{2\omega}\left[t' + \frac{\sin 2nt'}{2n}\right]_0^{2\pi} \\ &= \frac{\tau}{2},\end{aligned}$$

whereas if $m = n = 0$, the integral is clearly τ. Thus

$$\int_0^{\tau} \cos m\omega t \cos n\omega t \, dt = \begin{cases} 0 & \text{if } m \neq n, \\ \tau/2 & \text{if } m = n \neq 0, \\ \tau & \text{if } m = n = 0. \end{cases}$$

Similarly

$$\int_0^{\tau} \sin m\omega t \sin n\omega t \, dt = \begin{cases} 0 & \text{if } m \neq n, \\ \tau/2 & \text{if } m = n \neq 0 \end{cases}$$

and

$$\int_0^{\tau} \sin m\omega t \cos n\omega t \, dt = 0 \quad \text{for all } m, n.$$

The basis functions are therefore *orthogonal* over the range τ in that the product of two different members integrates to zero over this range. Because of the periodicity of the functions, it is not necessary to integrate from 0 to τ, and any time interval of length τ will do instead; for example,

$$\int_{t_0}^{t_0+\tau} \cos m\omega t \cos n\omega t \, dt = \int_0^{\tau} \cos m\omega t \cos n\omega t \, dt$$

for any t_0.

Euler's formulas

To determine the coefficients a_n in Eq. (4.3), multiply both sides by $\cos m\omega t$ and integrate from (say) 0 to τ. Since the series is finite, term-by-term integration is unquestionably valid and

$$\int_0^\tau f(t)\cos m\omega t\,dt = \frac{1}{2}a_0\int_0^\tau \cos m\omega t\,dt$$

$$+\sum_{n=1}^{\infty}\left(a_n\int_0^\tau \cos m\omega t\cos n\omega t\,dt + b_n\int_0^\tau \cos m\omega t\sin n\omega t\,dt\right).$$

If $1 \le m \le N$, the only nonzero integral on the right-hand side is that associated with the coefficient a_n with $n = m$, and thus

$$a_m = \frac{2}{\tau}\int_0^\tau f(t)\cos m\omega t\,dt.$$

On the other hand, if $m = 0$, the coefficient a_0 is picked out and

$$a_0 = \frac{2}{\tau}\int_0^\tau f(t)\,dt.$$

By taking the first term in Eq. (4.3) to be $a_0/2$, the formula for a_0 has been made a special case of that for a_m, and by changing the dummy index m back to n we have

$$a_n = \frac{2}{\tau}\int_\tau f(t)\cos n\omega t\,dt, \qquad n = 0, 1, 2, \ldots, N.$$

Similarly

$$b_n = \frac{2}{\tau}\int_\tau f(t)\sin n\omega t\,dt, \qquad n = 1, 2, 3, \ldots, N,$$

where the integration is over any interval in t of length $\tau = 2\pi/\omega$. These are known as *Euler's formulas* in recognition of Leonhard Euler.

Unfortunately the representation (4.3) with N finite is too restrictive to be of general use. Since each term on the right-hand side is continuous as a function of t, with all derivatives continuous, $f(t)$ and all its derivatives are likewise continuous. In contrast, most periodic functions of interest are less smooth or even piecewise-continuous like the square wave in Fig. 4.4, and it is impossible to represent these precisely using only a finite number of terms in the series. Expressed alternatively, the basis functions

$$1 \quad \text{and} \quad \cos n\omega t, \sin n\omega t \qquad \text{with } 1 \le n \le N$$

form a *complete* orthogonal set only if $N = \infty$. As a result we must allow the trigonometric series to be infinite, and the question of convergence now arises. Were it not for this, Fourier series would be a very sterile topic mathematically, but as it is, there are still matters relating to the convergence of the series that are the subject of research.

4.3 Fourier series

If $f(t)$ is periodic with period $\tau = 2\pi/\omega$, the Fourier-series representation of f is

$$\frac{1}{2}a_0 + \sum_{n=1}^{\infty} (a_n \cos n\omega t + b_n \sin n\omega t), \tag{4.5}$$

where the coefficients are given by Euler's formulas

$$a_n = \frac{2}{\tau}\int_\tau f(t)\cos n\omega t\,dt, \qquad n = 0, 1, 2, \ldots,$$

$$b_n = \frac{2}{\tau}\int_\tau f(t)\sin n\omega t\,dt, \qquad n = 1, 2, 3, \ldots. \tag{4.6}$$

Thus, the Fourier-series representation of $f(t)$ is a trigonometric series whose coefficients are as shown, and every trigonometric series with reasonable convergence properties is a Fourier series for some function f. The series is defined whenever the integrals in Eq. (4.6) have meaning, and this is certainly the case when $f(t)$ is continuous. But the integrals are also meaningful if $f(t)$ has finite jump discontinuities (i.e., if $f(t)$ is piecewise-continuous), and, as we shall see later, the Fourier-series representation of such a function does converge.

Since the coefficients are expressed as integrals over a τ range of t, it is only necessary to specify $f(t)$ in this range, and the Fourier series then represents the periodic extension of the given information for all t. Indeed, another application of Fourier series is the representation of functions defined over a finite range alone.

Example 1. Find the Fourier-series representation of

$$f(t) = \begin{cases} 1 & \text{for } 0 < t < 1, \\ 0 & \text{for } 1 < t < 2. \end{cases}$$

The periodic extension of this is the square wave plotted in Fig. 4.4. Its fundamental period is $\tau = 2$ so that $\omega = 2\pi/\tau = \pi$, and by choosing

(0, 2) for the range of integration, we have

$$\begin{aligned} a_n &= \int_0^2 f(t)\cos n\pi t\,dt \\ &= \int_0^1 \cos n\pi t\,dt \\ &= \left[\frac{\sin n\pi t}{n\pi}\right]_0^1 = 0, \qquad n \neq 0 \end{aligned}$$

with

$$a_0 = \int_0^1 dt = 1.$$

Also

$$\begin{aligned} b_n &= \int_0^1 \sin n\pi t\,dt \\ &= \left[-\frac{\cos n\pi t}{n\pi}\right]_0^1 = \frac{1-(-1)^n}{n\pi} \end{aligned}$$

since $\cos n\pi = (-1)^n$. The Fourier-series representation of $f(t)$ is therefore

$$\frac{1}{2} + \frac{1}{\pi}\sum_{n=1}^{\infty} \frac{1-(-1)^n}{n}\sin n\pi t, \tag{4.7}$$

the first few terms of which are

$$\frac{1}{2} + \frac{2}{\pi}\sin \pi t + \frac{2}{3\pi}\sin 3\pi t + \cdots.$$

If $S_n(t)$ denotes the partial sum of the Fourier series through terms in $\cos n\omega t$ and $\sin n\omega t$,

$$S_0(t) = \frac{1}{2}, \qquad S_1(t) = \frac{1}{2} + \frac{2}{\pi}\sin \pi t,$$

$$S_3(t) = \frac{1}{2} + \frac{2}{\pi}\sin \pi t + \frac{2}{3\pi}\sin 3\pi t, \quad \text{and so forth,}$$

and these should be compared with the analogous results obtained in Section 4.1 by inspection of the waveform. If t is an integer, (4.7) reduces to the constant term $\frac{1}{2}$, showing that at the jump discontinuities in f the Fourier-series representation converges to the average of the left- and right-hand limits of the function. By assuming that it converges to $f(t)$ at any point of continuity of f and putting $t = \frac{1}{2}$, we obtain

$$1 = \frac{1}{2} + \frac{2}{\pi}\left(1 - \frac{1}{3} + \frac{1}{5} - \frac{1}{7} + \cdots\right)$$

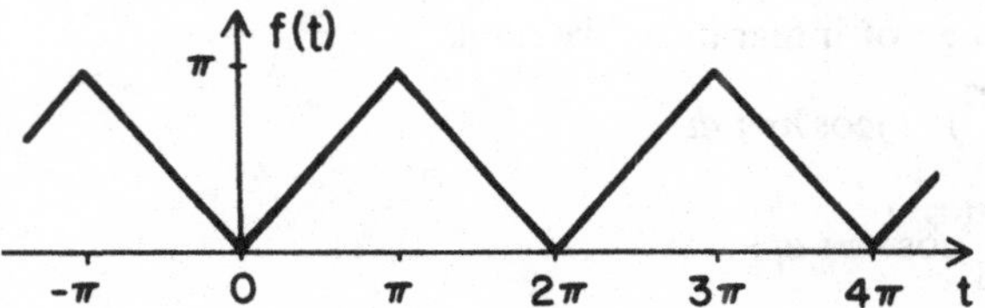

Fig. 4.5: Triangular wave.

or

$$1-\frac{1}{3}+\frac{1}{5}-\frac{1}{7}+\cdots=\frac{\pi}{4}.$$

This famous result was discovered by the German mathematician Gottfried Wilhelm Leibniz (1646–1716) using geometrical considerations, and shows how series with constant terms can be summed by evaluating Fourier series at specific points.

Example 2. Find the Fourier-series representation of

$$f(t)=|t| \qquad \text{for } |t|<\pi.$$

The periodic extension of this is the triangular wave shown in Fig. 4.5. Its fundamental period is $\tau=2\pi$, implying $\omega=2\pi/\tau=1$. For convenience we choose the range of integration to be $(-\pi,\pi)$ and by recognizing that $|t|=-t$ if $t<0$, we obtain

$$\begin{aligned}
a_n &= \frac{1}{\pi}\int_{-\pi}^{\pi} f(t)\cos nt\,dt \\
&= \frac{1}{\pi}\left\{\int_{-\pi}^{0}(-t)\cos nt\,dt+\int_0^{\pi} t\cos nt\,dt\right\} \\
&= \frac{2}{\pi}\int_0^{\pi} t\cos nt\,dt \\
&= \frac{2}{\pi}\left\{\left[\frac{t\sin nt}{n}\right]_0^{\pi}-\frac{1}{n}\int_0^{\pi}\sin nt\,dt\right\} \qquad n\neq 0 \\
&= \frac{2}{\pi}\left[\frac{t\sin nt}{n}+\frac{\cos nt}{n^2}\right]_0^{\pi}=-\frac{2}{\pi n^2}\{1-(-1)^n\}
\end{aligned}$$

with

$$a_0=\frac{2}{\pi}\int_0^{\pi} t\,dt=\pi.$$

Also

$$b_n=\frac{1}{\pi}\left\{\int_{-\pi}^{0}(-t)\sin nt\,dt+\int_0^{\pi} t\sin nt\,dt\right\}=0.$$

The Fourier-series representation of $f(t)$ is therefore

$$\frac{\pi}{2}-\frac{2}{\pi}\sum_{n=1}^{\infty}\frac{1-(-1)^n}{n^2}\cos nt, \tag{4.8}$$

the first few terms of which are

$$\frac{\pi}{2}-\frac{4}{\pi}\cos t-\frac{4}{9\pi}\cos 3t-\cdots .$$

As evident from these examples, the determination of the Fourier-series representation of a function $f(t)$ is a straightforward task whose main feature is the evaluation of certain integrals involving sines and cosines. Unfortunately, it is easy to make a mistake even prior to the integration, and in order to reduce the possibility of error, it is suggested that the task be broken up as follows:

(i) Based on the given information, determine the *periodic* function $f(t)$ and plot this as a function of t.

(ii) Determine the fundamental period τ in terms of which $\omega = 2\pi/\tau$. Though it is sufficient to allow τ to be any period of $f(t)$, the analytical work is increased if other than the fundamental period is chosen.

(iii) Select the interval of length τ for the integration. If $f(t)$ is specified over a finite range only (so we are actually seeking the Fourier-series representation of the periodic continuation of the given function), it is natural, but not essential, to choose this as the interval. A wise choice of interval can simplify the analysis, and the functional form of $f(t)$ may depend on the choice. If, in Example 2, we had chosen the interval $(0, 2\pi)$, the required definition of $f(t)$ would have been

$$f(t)=\begin{cases} t & \text{for } 0<t<\pi, \\ 2\pi-t & \text{for } \pi<t<2\pi. \end{cases}$$

(iv) Evaluate Euler's formulas for the coefficients. This will often involve integration by parts, and although the formulas are valid for all nonnegative integers n, the integration may require that special values of n be excluded and considered separately. The excluded cases are those for which the denominator vanishes, for example, $n = 0$ in the preceding examples: Division by zero is not permitted. It is suggested that the excluded cases be noted as they arise and treated separately by going back to the line previous to the notation.

(v) State the resulting Fourier-series representation of $f(t)$. Since there is some variation in the forms used throughout the literature (some texts, for example, omit the factor $\frac{1}{2}$ multiplying a_0 in Eq. (4.5)), do not leave the reader to guess the form that you have employed.

Exercises

Find the Fourier-series representations of the following $f(t)$, periodically extended where necessary.

*1. $f(t) = \begin{cases} 1 & \text{for } 0 < t < 1, \\ 0 & \text{for } 1 < t < 4. \end{cases}$

2. $f(t) = \begin{cases} 1 & \text{for } 0 < t < \pi, \\ -1 & \text{for } \pi < t < 2\pi, \\ 0 & \text{for } 2\pi < t < 4\pi. \end{cases}$

3. $f(t) = \begin{cases} t & \text{for } 0 < t < 1, \\ 0 & \text{for } 1 < t < 2. \end{cases}$ 4. $f(t) = \begin{cases} t & \text{for } 0 < t < 1, \\ 1 & \text{for } 1 < t < 2. \end{cases}$

5. $f(t) = \begin{cases} t & \text{for } 0 < t < 1, \\ 1 & \text{for } 1 < t < 2, \\ 0 & \text{for } 2 < t < 4. \end{cases}$

6. $f(t) = \begin{cases} \sin t & \text{for } 0 < t < \pi/2, \\ 0 & \text{for } \pi/2 < t < 2\pi. \end{cases}$

*7. $f(t) = \begin{cases} \cos t & \text{for } 0 < t < \pi/2, \\ 0 & \text{for } \pi/2 < t < \pi. \end{cases}$

8. $f(t) = \begin{cases} \sin t & \text{for } 0 < t < \pi, \\ 0 & \text{for } \pi < t < 2\pi. \end{cases}$

9. $f(t) = \sin^2 t$ (Think!). 10. $f(t) = \sin t|\sin t|$.

4.4 Symmetry conditions

In the examples in the previous section, some of the Fourier coefficients are zero. This is due to the symmetry properties of the periodic functions. When the function has certain symmetries, inspection can suffice to tell which terms are absent from the Fourier-series representation and to simplify the expressions for the remaining coefficients. We shall discuss the symmetry properties under two headings.

Even and odd functions

(i) If $f(t)$ is such that

$$f(-t) = f(t) \tag{4.9}$$

for all t, $f(t)$ is said to be an *even* or *symmetric* (about $t = 0$) function. Of course, an even function of period τ is symmetric not only with respect to the vertical axis through $t = 0$ but also with respect to the vertical lines $t = \pm m\tau$, $m = 1, 2, 3, \ldots$.

To determine the Fourier coefficients, it is convenient to choose the range of integration to be $(-\tau/2, \tau/2)$. Then

$$\begin{aligned} b_n &= \frac{2}{\tau}\int_{-\tau/2}^{\tau/2} f(t)\sin n\omega t\, dt \\ &= \frac{2}{\tau}\left\{\int_{-\tau/2}^{0} f(t)\sin n\omega t\, dt + \int_0^{\tau/2} f(t)\sin n\omega t\, dt\right\} \\ &= \frac{2}{\tau}\left\{-\int_0^{\tau/2} f(-t')\sin n\omega t'\, dt' + \int_0^{\tau/2} f(t)\sin n\omega t\, dt\right\} \\ &= 0 \end{aligned}$$

by virtue of Eq. (4.9). Thus, all of the sine terms are absent from the representation, which now takes the form

$$\frac{1}{2}a_0 + \sum_{n=1}^{\infty} a_n \cos n\omega t. \tag{4.10}$$

This is called a *Fourier cosine series*. The coefficients a_n are defined as before, namely,

$$a_n = \frac{2}{\tau}\int_{-\tau/2}^{\tau/2} f(t)\cos n\omega t\, dt, \qquad n = 0, 1, 2, \ldots,$$

but we can again use Eq. (4.9) to write

$$a_n = \frac{4}{\tau}\int_0^{\tau/2} f(t)\cos n\omega t\, dt, \qquad n = 0, 1, 2, \ldots, \tag{4.11}$$

showing that for an even function a knowledge of $f(t)$ over a half-period (or half-range) suffices. For this reason, (4.10) with coefficients (4.11) is often called a *half-range expansion*. Since (4.10) is an even function of t, a representation in the form of a Fourier cosine series requires that $f(t)$ is even.

(ii) If $f(t)$ is such that

$$f(-t) = -f(t) \tag{4.12}$$

for all t, $f(t)$ is said to be an odd or antisymmetric (about $t = 0$) function. Of course, an odd function of period τ is also antisymmetric with respect to all the vertical lines $t = \pm m\tau$, $m = 1, 2, 3, \ldots$.

By a process similar to that used above, it follows immediately that

$$a_n = 0, \qquad n = 0, 1, 2, \ldots .$$

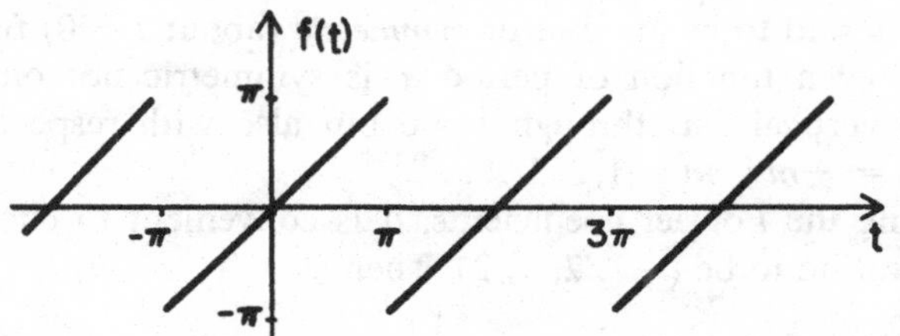

Fig. 4.6: The periodic function having $f(t) = t$ for $|t| < \pi$.

Thus, all cosine terms including the constant are absent from the representation, which now has the form

$$\sum_{n=1}^{\infty} b_n \sin n\omega t. \tag{4.13}$$

This is called a *Fourier sine series*. The coefficients b_n are defined as before, but because of the antisymmetry we can also write

$$b_n = \frac{4}{\tau}\int_0^{\tau/2} f(t)\sin n\omega t\, dt, \tag{4.14}$$

which requires a knowledge of $f(t)$ over half the period or range only. The series (4.13) with coefficients (4.14) is the second example of a half-range expansion. Since (4.13) is an odd function of t, a representation in the form of a Fourier sine series demands that $f(t)$ be odd.

Example 1. Find the Fourier sine series corresponding to

$$f(t) = t, \qquad 0 < t < \pi.$$

The problem is that of finding the Fourier-series representation of the periodic continuation of the corresponding odd function, and the resulting periodic function is shown in Fig. 4.6. Since $\tau = 2\pi$, implying $\omega = 1$,

$$\begin{aligned} b_n &= \frac{1}{\pi}\int_{-\pi}^{\pi} f(t)\sin nt\, dt \\ &= \frac{2}{\pi}\int_0^{\pi} t\sin nt\, dt \qquad \text{(c.f., Eq. (4.14))} \\ &= \frac{2}{\pi}\left[-\frac{t\cos nt}{n} + \frac{\sin nt}{n^2}\right]_0^{\pi} \end{aligned}$$

using integration by parts. Hence

$$b_n = -2\frac{(-1)^n}{n}$$

and the Fourier-sine-series representation of $f(t)$ is

$$-2\sum_{n=1}^{\infty}\frac{(-1)^n}{n}\sin nt. \tag{4.15}$$

Example 2. Find the Fourier cosine series corresponding to

$$f(t) = t, \qquad 0 < t < \pi.$$

Since the function must be even, the task is to find the Fourier-series representation of the periodic continuation of $f(t) = |t|$ for $|t| < \pi$. This is identical to the problem treated in Example 2 of Section 4.3 (see Fig. 4.5), and the representation is

$$\frac{\pi}{2} - \frac{2}{\pi}\sum_{n=1}^{\infty}\frac{1-(-1)^n}{n^2}\cos nt. \tag{4.16}$$

Most periodic functions are neither even nor odd, but any function $f(t)$ can be expressed as the sum of even and odd functions. Clearly

$$f(t) = \tfrac{1}{2}\{f(t) + f(-t)\} + \tfrac{1}{2}\{f(t) - f(-t)\} \tag{4.17}$$

and the first term on the right-hand side is an even function, whereas the second term is odd. In the Fourier-series representation of a general function $f(t)$, the constant and cosine terms comprising a Fourier cosine series take care of the first (even) part of Eq. (4.17), and the sine terms comprising a Fourier sine series take care of the second (odd) part.

Even and odd harmonics

There is a further type of symmetry that can be exploited and that serves to eliminate all terms from the Fourier-series representation having even values of n (including $n = 0$) or odd values. The terms corresponding to the even values are referred to as the even harmonics, while those corresponding to odd values are the odd harmonics.

(i) If $f(t)$ is such that

$$f\left(t \pm \frac{\tau}{2}\right) = f(t), \tag{4.18}$$

where τ is the period, then only the even harmonics are present.

To prove that the odd harmonics are absent, it is sufficient to consider the coefficients a_n for n odd. We have

$$\begin{aligned}
a_n &= \frac{2}{\tau}\int_0^{\tau} f(t)\cos n\omega t\, dt \\
&= \frac{2}{\tau}\left\{\int_0^{\tau/2} f(t)\cos n\omega t\, dt + \int_{\tau/2}^{\tau} f(t)\cos n\omega t\, dt\right\} \\
&= \frac{2}{\tau}\left\{\int_0^{\tau/2} f(t)\cos n\omega t\, dt + \int_0^{\tau/2} f\left(t' + \frac{\tau}{2}\right)\cos n\omega\left(t' + \frac{\tau}{2}\right) dt'\right\} \\
&= \frac{2}{\tau}\left\{\int_0^{\tau/2} f(t)\cos n\omega t\, dt - \int_0^{\tau/2} f\left(t' + \frac{\tau}{2}\right)\cos n\omega t'\, dt'\right\}.
\end{aligned}$$

Hence, from Eq. (4.18), $a_n = 0$. Similarly $b_n = 0$ for n odd, whereas for n even

$$\begin{aligned}
a_n &= \frac{4}{\tau}\int_0^{\tau/2} f(t)\cos n\omega t\, dt, \\
b_n &= \frac{4}{\tau}\int_0^{\tau/2} f(t)\sin n\omega t\, dt. \qquad (4.19)
\end{aligned}$$

As in the case of an even or odd function (about $t = 0$), it is now sufficient to integrate over half the period.

Example 1. Find the Fourier-series representation of the rectified sine wave

$$f(t) = |\sin t|.$$

The function is shown in Fig. 4.7, and to exploit its evenness, the period is taken to be $\tau = 2\pi$, implying $\omega = 1$. With this choice of τ, $f(t)$ is symmetric about $t = \pm\tau/4$ and therefore satisfies Eq. (4.18), but if we proceed using only the fact that $f(t)$ is even,

$$\begin{aligned}
a_n &= \frac{1}{\pi}\int_{-\pi}^{\pi} |\sin t|\cos nt\, dt \\
&= \frac{2}{\pi}\int_0^{\pi} \sin t\cos nt\, dt \\
&= \frac{1}{\pi}\int_0^{\pi} \{\sin(1-n)t + \sin(1+n)t\}\, dt \\
&= \frac{1}{\pi}\left[-\frac{\cos(1-n)t}{1-n} - \frac{\cos(1+n)t}{1+n}\right]_0^{\pi} \qquad n \neq 1 \\
&= -\frac{2}{\pi}\frac{1+(-1)^n}{n^2-1}
\end{aligned}$$

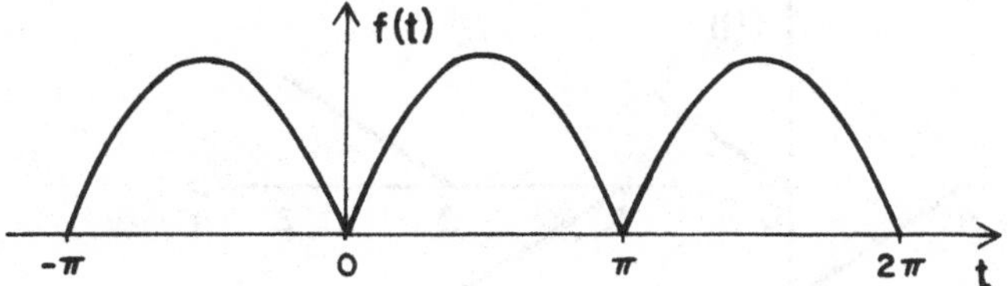

Fig. 4.7: Rectified sine wave $|\sin t|$.

with

$$a_1 = \frac{1}{\pi}\int_0^{\pi} \sin 2t\, dt = 0.$$

The Fourier-series representation is therefore

$$\frac{2}{\pi} - \frac{2}{\pi}\sum_{n=2}^{\infty} \frac{1+(-1)^n}{n^2-1}\cos nt, \tag{4.20}$$

the first few terms of which are

$$\frac{2}{\pi}\left\{1 - \frac{2}{3}\cos 2t - \frac{2}{15}\cos 4t - \cdots\right\}.$$

As expected, only the even harmonics are present.

To understand the effect of this type of symmetry, it is sufficient to consider the behavior of the sines and cosines. For the odd harmonics the sines are symmetric about $t = \pm\tau/4$, whereas the cosines are antisymmetric. For the even harmonics the behavior is reversed, and by invoking these symmetries it can be shown that

> a necessary and sufficient condition for a Fourier series to contain only even harmonics is that the actual fundamental period of the function is one-half of the period used in the derivation of the series.

In the preceding example the fundamental period of $f(t)$ is π.

(ii) If $f(t)$ is such that

$$f\left(t \pm \frac{\tau}{2}\right) = -f(t), \tag{4.21}$$

where τ is the period, then only the odd harmonics are present.

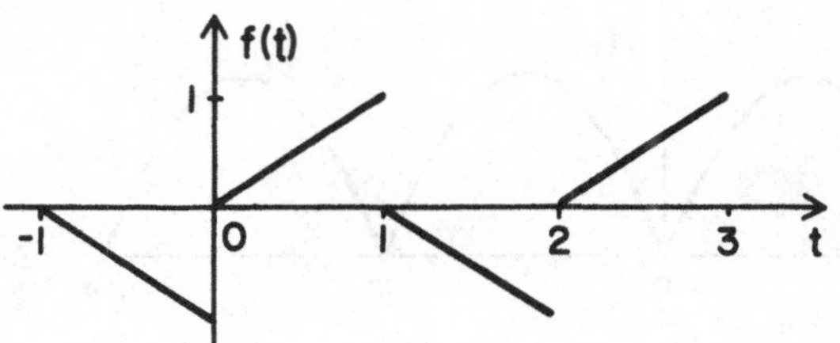

Fig. 4.8: A periodic function having $f(t \pm 1) = -f(t)$.

The proof is similar to that given above; thus, for n even,

$$a_n = b_n = 0$$

whereas for n odd,

$$a_n = \frac{4}{\tau}\int_0^{\tau/2} f(t)\cos n\omega t \, dt,$$

$$b_n = \frac{4}{\tau}\int_0^{\tau/2} f(t)\sin n\omega t \, dt. \tag{4.22}$$

Example 2. Find the Fourier-series representation of

$$f(t) = \begin{cases} t & \text{for } 0 < t < 1, \\ -(t-1) & \text{for } 1 < t < 2. \end{cases}$$

The periodic function is shown in Fig. 4.8, and since $\tau = 2$, implying $\omega = \pi$, Eq. (4.21) is satisfied. If we proceed directly without invoking this symmetry and choose the range of integration to be (0, 2), we have

$$\begin{aligned} a_n &= \int_0^2 f(t)\cos n\pi t \, dt \\ &= \int_0^1 t\cos n\pi t \, dt - \int_1^2 (t-1)\cos n\pi t \, dt \\ &= \int_0^1 t\cos n\pi t \, dt - \int_0^1 t'\cos n\pi(t'+1) \, dt' \\ &= \{1-(-1)^n\}\int_0^1 t\cos n\pi t \, dt \\ &= \{1-(-1)^n\}\left[\frac{t\sin n\pi t}{n\pi} + \frac{\cos n\pi t}{(n\pi)^2}\right]_0^1 \qquad n \neq 0 \\ &= -\left\{\frac{1-(-1)^n}{n\pi}\right\}^2 \end{aligned}$$

with

$$a_0 = 0.$$

Also

$$\begin{aligned} b_n &= \int_0^1 t \sin n\pi t \, dt - \int_0^1 t' \sin n\pi(t' + 1) \, dt' \\ &= \{1 - (-1)^n\} \int_0^1 t \sin n\pi t \, dt \\ &= \{1 - (-1)^n\} \left[-\frac{t \cos n\pi t}{n\pi} + \frac{\sin n\pi t}{(n\pi)^2} \right]_0^1 \\ &= \frac{1 - (-1)^n}{n\pi}. \end{aligned}$$

The Fourier-series representation of $f(t)$ is therefore

$$\frac{1}{\pi} \sum_{n=1}^{\infty} \frac{1 - (-1)^n}{n} \left\{ -\frac{1 - (-1)^n}{n\pi} \cos n\pi t + \sin n\pi t \right\}, \tag{4.23}$$

the first few terms of which are

$$-\frac{4}{\pi^2} \left\{ \cos \pi t + \frac{1}{9} \cos 3\pi t + \cdots \right\} + \frac{2}{\pi} \left\{ \sin \pi t + \frac{1}{3} \sin 3\pi t + \cdots \right\}.$$

As expected, only the odd harmonics are present.

When analyzing a periodic function it is always desirable to check first to see if any symmetries exist. When they do, it may be possible to simplify the calculation of the Fourier coefficients and to avoid the determination of some of them entirely. Sometimes a function does not satisfy any of the symmetry conditions we have given, but can be made to do so by a simple shift of the horizontal axis. An example is provided by the square wave in Fig. 4.4. As it stands, it is neither an even nor an odd function, but apart from the constant term its Fourier series is a sine series with only odd harmonic present. The reason becomes clear when we consider $f(t) - \frac{1}{2}$, which satisfies Eq. (4.21) as well as Eq. (4.12). A parallel shift of the horizontal axis changes only the average value of the function, and symmetry properties should be examined after the horizontal axis has been shifted to the position of the average value. A parallel shift of the vertical axis amounts to a shift in the origin of t, and whereas this may change the sine or cosine structure of the Fourier-series representation, it does not change the harmonic content. This accords with the fact that the harmonic content of a periodic function is a physical attribute that is unaffected by the arbitrary choice of the origin of time.

For the examples and exercises in this chapter where the functions have relatively simple forms, the only symmetry that has a significant effect on the determination of the Fourier coefficients is evenness or oddness of the function. In contrast to the harmonic property, symmetry about $t = 0$ eliminates an entire set of coefficients.

4.5 Mathematical properties

The mere fact that a Fourier series has been written down does not ensure its convergence or, if convergent, that its sum is the function $f(t)$ used to generate the series. We have so far consciously avoided equating a function to its Fourier-series representation, but if we are to employ these series in the solution of physical problems, it is necessary to determine the conditions under which a periodic function can be replaced by (or equated to) its series. That this is not entirely a trivial matter can be seen from the Examples 1 and 2 in Section 4.3. In each case the Fourier series is demonstrably convergent and defines a function for all t, $-\infty < t < \infty$, but for the square wave the function $f(t)$ used to establish the series is not even defined for integer t.

Before discussing the convergence of Fourier series it is necessary to define the functions of particular concern and to review some basic notions about the convergence of series in general.

Piecewise-continuous and -smooth functions

In Section 2.2 we introduced the idea of right- and left-hand limits. As applied to a function $f(t)$,

$$\text{right-hand limit} = \lim_{\Delta t \to 0+} f(t + \Delta t) = f(t+),$$

$$\text{left-hand limit} = \lim_{\Delta t \to 0-} f(t + \Delta t) = f(t-).$$

If t is a point of continuity of $f(t)$, the right- and left-hand limits exist and are equal (i.e., $f(t+) = f(t-)$), which we write simply as $f(t)$.

In the same way we can define the right- and left-hand derivatives of $f(t)$:

$$\text{right-hand derivative} = \lim_{\Delta t \to 0+} \frac{f(t + \Delta t) - f(t+)}{\Delta t} = f'(t+),$$

$$\text{left-hand derivative} = \lim_{\Delta t \to 0-} \frac{f(t + \Delta t) - f(t-)}{\Delta t} = f'(t-).$$

At any point where $f(t)$ has an ordinary derivative (i.e., is differentiable), the two limits exist and are equal, and define the ordinary derivative

$f'(t)$, namely,

$$f'(t+) = f'(t-) = f'(t).$$

A function that is differentiable at t is necessarily continuous at t, and, because of this, it is possible for a function to have one-sided derivatives that are equal without having an ordinary derivative. An example is the unit-step function $f(t) = u(t)$ whose one-sided derivatives at $t = 0$ exist and are equal, but $f'(0)$ does not exist.

As previously defined, a function $f(t)$ is *piecewise-continuous* in $a \leq t \leq b$ if it is continuous apart from a finite number of points $t = t_n$ where it has finite jump discontinuities. In each subinterval $t_{n-1} \leq t \leq t_n$ the one-sided limits exist at the endpoints, and they exist and are equal at all interior points. If, in addition, $f(t)$ is differentiable in each subinterval, that is, the one-sided derivatives exist at the endpoints, and they exist and are equal at all interior points, $f(t)$ is said to be *piecewise-smooth*. This is obviously a more stringent condition than piecewise continuity. A piecewise-smooth function is necessarily piecewise-continuous and its derivative is also piecewise continuous.

Convergence concepts

The series of terms

$$c_1 + c_2 + c_3 + \cdots = \sum_{n=1}^{\infty} c_n$$

converges if the sequence of partial sums S_n converges, that is, if

$$\lim_{n \to \infty} \{S_n\} = S < \infty,$$

where

$$S_n = c_1 + c_2 + \cdots + c_n.$$

Expressed alternatively, the series has the sum S if, given $\varepsilon > 0$, there exists $N = N(\varepsilon)$ such that

$$|S_n - S| < \varepsilon$$

for all $n > N$. Both statements are consistent with a numerical approach in which ever-increasing numbers of terms are summed until the answer no longer changes to the accuracy desired. Of course, if this stage is never reached, one is forced to the conclusion that the series diverges. Since

$$c_n = S_n - S_{n-1},$$

convergence of the series requires that

$$\lim_{n\to\infty} c_n = \lim_{n\to\infty} \{S_n\} - \lim_{n\to\infty} \{S_{n-1}\} = 0.$$

The vanishing of the individual terms as $n \to \infty$ is a necessary but not sufficient condition for convergence.

If the series $\Sigma_{n=1}^{\infty}|c_n|$ is also convergent, the series $\Sigma_{n=1}^{\infty}c_n$ is said to be *absolutely convergent*. This is a stronger requirement, which rules out the possibility of convergence by virtue of a sign alternation, and is necessary if the sum is to be unaffected by a rearrangement of terms.

Suppose that the terms are functions of a variable t. If the series $\Sigma_{n=1}^{\infty}c_n(t)$ is convergent, the sum S is also a function of t, and for any $\varepsilon > 0$ there exists an integer N such that

$$|S_n(t) - S(t)| < \varepsilon$$

for all $n > N$. In general N will be a function of t as well as ε, but if it is possible to determine an N that is independent of t, the series $\Sigma_{n=1}^{\infty}c_n(t)$ is said to be *uniformly convergent* to $S(t)$. If the terms $c_n(t)$ are continuous functions of t, the sum $S(t)$ is itself continuous.

Fourier theorems

Theorems that give conditions under which a Fourier series corresponding to a function converges to that function are generally referred to as *Fourier theorems*. There are a number of such theorems and these differ in several particulars. One reason for the differences is that the conditions are all sufficient and not necessary: Necessary conditions for the convergence of a Fourier series are not yet known. Another is the distinction between the Fourier-series representation of a periodic function defined for $-\infty < t < \infty$ and the representation of the periodic continuation (or extension) of a function defined in (say) $-\tau/2 \leq t \leq \tau/2$. If, in the latter case, $f(t)$ is continuous in the interval but $f(-\tau/2) \neq f(\tau/2)$, the periodic function has jump discontinuities at the endpoints of the range and is only piecewise-continuous in $-\infty < t < \infty$.

A fundamental theorem is as follows. (Like all of the results in this section it is presented without proof, and to see the way in which the theorems are established, the reader is referred to the more mathematically oriented texts listed at the end of the chapter.)

> If $f(t)$ is a periodic piecewise-continuous function with period τ, the Fourier series (4.5) with coefficients (4.6) converges to
>
> $\frac{1}{2}\{f(t+) + f(t-)\}$

at every point where $f(t)$ has a right-hand and a left-hand derivative.

Thus, the series converges everywhere if $f(t)$ is piecewise-smooth. An alternative statement is provided by the *Dirichlet conditions*, named after the German mathematician Peter Gustav Lejeune Dirichlet (1805–59): $f(t)$ is bounded, with only a finite number of ordinary (or jump) discontinuities and a finite number of maxima and minima in any finite range. We remark that a function which is piecewise-continuous in (a, b) is necessarily bounded in (a, b).

At any point where $f(t)$ is continuous

$$f(t+) = f(t-) = f(t),$$

and the average of the one-sided limits is the value of the function itself. If the one-sided derivatives exist there, the series converges to $f(t)$. Theoretically at least, the existence of these derivatives is not required for all t but only where the representation is used. The convergence of a Fourier series is determined by the *local behavior* of $f(t)$. At points where the one-sided derivatives do not exist, the convergence of the series is not ensured by the theorem; where the derivatives do exist, the convergence is guaranteed and the series provides a representation.

If $f(t)$ is piecewise-continuous in $-\infty < t < \infty$, or continuous in (let us say) $-\tau/2 \leq t \leq \tau/2$ with $f(-\tau/2) \neq f(\tau/2)$, it is strictly speaking not correct to equate $f(t)$ to its Fourier series even if the one-sided derivatives exist everywhere. At a point of jump discontinuity $f(t)$ is undefined, whereas the Fourier series converges to the average of the one-sided limits. Some texts draw attention to this by using a symbol other than the equality sign, but since the discrepancy at these isolated points is of no importance for the applications we have in mind, we shall henceforth equate a function to its Fourier series whenever the series converges.

The sum of a uniformly convergent series of continuous functions is itself continuous, implying that the Fourier-series representation of a piecewise-smooth function does not necessarily converge uniformly. However, if $f(t)$ is also continuous (which allows it to have "corners"), the series converges uniformly and absolutely for all t, and even when $f(t)$ is only piecewise-continuous, the convergence is still uniform in each closed interval containing no jump points. Conversely, if a Fourier series converges uniformly for all t, it represents a continuous periodic function; two Fourier series which converge uniformly for all t and have the same sum have the same coefficients; that is, the Fourier series is a *unique* representation.

If $f(t)$ is piecewise-smooth, its Fourier series can be differentiated term by term and converges to $f'(t)$ at each point where $f(t)$ has an ordinary derivative; that is, if

$$f(t) = \frac{1}{2}a_0 + \sum_{n=1}^{\infty} (a_n \cos n\omega t + b_n \sin n\omega t),$$

then

$$f'(t) = \omega \sum_{n=1}^{\infty} n(b_n \cos n\omega t - a_n \sin n\omega t).$$

On the other hand, if $f(t)$ is merely piecewise-continuous, its Fourier series can be integrated term by term regardless of whether the series converges or not, and

$$\frac{1}{2}a_0\left(t + \frac{\tau}{2}\right) + \frac{1}{\omega}\sum_{n=1}^{\infty} \frac{1}{n}\left(a_n \sin n\omega t - b_n\left[\cos n\omega t - (-1)^n\right]\right)$$

is the Fourier-series representation of the periodic continuation of

$$\int_{-\tau/2}^{t} f(t')\,dt', \qquad -\frac{\tau}{2} \le t \le \frac{\tau}{2}.$$

Rate of convergence

The rate of convergence of a Fourier series is determined by its coefficients and the manner in which they decrease with increasing n. As shown above, term-by-term differentiation of the series multiplies the coefficients by $n\omega$ and thereby reduces the convergence, possibly to the extent of making the series diverge. In contrast, integration divides the coefficients by $n\omega$ and improves the convergence. Integration is a smoothing operation, whereas differentiation reduces the smoothness of the function, sharpening the peaks and even creating discontinuities where none existed before. It therefore comes as no surprise that the rate of convergence increases with the smoothness of the function.

In Example 1 of Section 4.3 the coefficients drop off as $1/n$ and the rate of convergence of the Fourier series (4.7) is a function of t. It is slowest near a jump discontinuity in $f(t)$, but even here the series does converge, albeit slowly, reflecting the difficulty that the beautifully smooth sines have in modeling the jump. In Example 2 of the same section, $f(t)$ represents a triangular wave and is a continuous function with a discontinuous first derivative. Its Fourier series is given in Eq. (4.8) and the coefficients drop off as $1/n^2$. The convergence is more rapid, and this is reasonable because a triangular wave is smoother than a square wave.

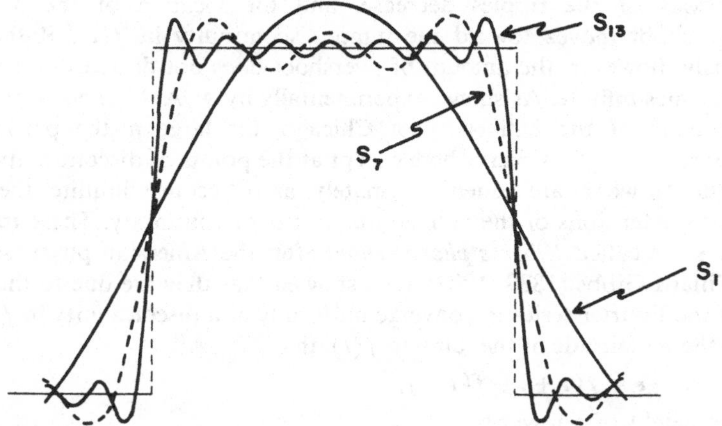

Fig. 4.9: The approximations provided by the partial sums S_1, S_7, and S_{13}.

Observations such as these can be summarized in certain general statements about the rate of decrease of the Fourier coefficients when n is large.

(i) If $f(t)$ is only piecewise-continuous, the coefficients decrease as $1/n$.
(ii) If $f(t)$ is continuous everywhere but its first derivative has discontinuities, the coefficients decrease as $1/n^2$.
(iii) If $f(t)$ and its derivatives through the $(k-1)$th order are continuous everywhere but its kth derivative has discontinuities, the coefficients decrease as $1/n^{k+1}$.

The rate of convergence of a Fourier series has important implications for the generation of a periodic function, and if the function required is only piecewise-continuous, it may be necessary to include a large number of terms to obtain an adequate approximation to the function. Figure 4.9 shows the square wave of Fig. 4.4 along with the approximations provided by a number of the partial sums derived from the series (4.7). The approximations have several features in common. Each is a continuous function of t, with all derivatives continuous, and thus the maximum rate of increase (or decrease) is finite, even near a jump discontinuity in $f(t)$. At a discontinuity $S_n = \frac{1}{2}$ for all n and the rate of change of $S_n(t)$ at this point increases with increasing n. Each approximation initially overshoots the desired level of 1 or 0 and then oscillates with an amplitude that decreases toward the middle. As n increases, the amplitudes and the periods of the ripples decrease, and the location of the maximum overshoot moves toward the jump discontinuity in $f(t)$. Rather curiously, however, the amount of overshoot does not decrease to zero as n becomes infinite. As shown experimentally by A. A. Michelson and S. W. Stratton of the University of Chicago, for large n the partial sums approximate $f(t)$ everywhere except at the points of discontinuity, where little "towers" are added. Ultimately, as n becomes infinite, the towers form extensions of the vertical line at the discontinuity. These irregularities are called *Gibbs's phenomenon*, after the American physicist Josiah Willard Gibbs (1839–1903), who showed that they are due to the failure of the Fourier series to converge uniformly at a discontinuity in $f(t)$. If ε is the magnitude of the jump in $f(t)$, that is

$$\varepsilon = f(t+) - f(t-),$$

the height of a tower is

$$\frac{\varepsilon}{\pi}\int_{\pi}^{\infty}\frac{\sin x}{x}\,dx = 0.089\varepsilon.$$

The towers can be suppressed by using suitable smoothing or convergence factors.

4.6 Physical properties and applications

The Fourier-series representation of a periodic function is vital to the solution of many problems, one of which is the determination of the steady-state response of a system. This is the reason for considering the series at this time; also, a Fourier series provides information about the nature of the periodic function itself.

Frequency spectrum

If τ is the fundamental period of a periodic function $f(t)$ whose Fourier series converges, we have

$$f(t) = \frac{1}{2}a_0 + \sum_{n=1}^{\infty}(a_n\cos n\omega t + b_n\sin n\omega t) \tag{4.24}$$

with

$$a_n = \frac{2}{\tau}\int_{\tau} f(t)\cos n\omega t\,dt, \qquad n = 0, 1, 2, \ldots,$$

$$b_n = \frac{2}{\tau}\int_{\tau} f(t)\sin n\omega t\,dt, \qquad n = 1, 2, 3, \ldots,$$

where $\omega = 2\pi/\tau$ and the integration is over an interval in t of length τ.

Each term in the Fourier series (4.24) has a physical interpretation. The leading term is

$$\frac{1}{2}a_0 = \frac{1}{\tau}\int_\tau f(t)\,dt,$$

which is simply the average value of $f(t)$. This is the time-independent part of the representation and is the dc (for direct current) or bias-level term. It can be changed at will by moving the horizontal (or t) axis up or down without affecting the shape of $f(t)$. The other terms in Eq. (4.24) are functions of t. As noted earlier, $\cos\omega t$ and $\sin\omega t$ are simple-harmonic or single-frequency oscillations whose angular frequency ω is the fundamental angular frequency $2\pi/\tau$ of $f(t)$. Each differs from the other by a $\tau/4$ shift in the origin of t, and the cosine and sine are often referred to as the in-phase and phase-quadrature components, respectively. The terminology comes from the use of a complex phasor

$$e^{i\omega t} = \cos\omega t + i\sin\omega t$$

and the interpretation of i as a $\pi/2$ rotation in the complex plane. Thus, $a_1\cos\omega t$ and $b_1\sin\omega t$ are the in-phase and phase-quadrature components of amplitudes a_1 and b_1, respectively, with frequency equal to the fundamental angular frequency. Similarly, $a_2\cos 2\omega t$ and $b_2\sin 2\omega t$ are in-phase and phase-quadrature components of amplitudes a_2 and b_2, respectively, with frequency twice the fundamental, and so on. In particular, the frequency of the nth terms is n times the fundamental. Thus, apart from the dc term, the frequencies involved in the representation of $f(t)$ are all integer multiples of the fundamental angular frequency.

The preceding information can be displayed in the form of plots of the coefficients a_n and b_n as functions of the angular frequency, as shown in Fig. 4.10. These constitute the *frequency spectrum* of $f(t)$, and since only the frequencies $n\omega$ occur, the spectrum consists of discrete lines spaced ω apart. The spectrum of a periodic function is therefore a *discrete* or *line spectrum*. For any periodic function the computation of the Fourier coefficients a_n and b_n is, in effect, the determination of the frequency spectrum of the function. Figure 4.10 follows immediately from Eq. (4.24) and, conversely, the figure contains all of the information necessary to construct the Fourier-series representation of the function.

Synthesis of $f(t)$

The computation of the a_n and b_n is a problem in analysis in which we are seeking the decomposition of a periodic function into its elementary parts. The resulting Fourier series then constitutes a synthesis of $f(t)$,

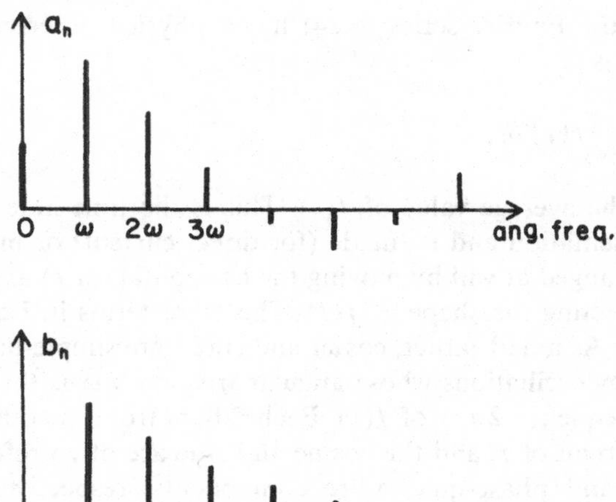

Fig. 4.10: Typical frequency spectrum.

and one reason for needing this is to generate a desired $f(t)$. The function might, for example, be a radar signal consisting of a train of pulses having a particular shape to help discriminate against unwanted targets. Its fundamental frequency is clearly the pulse-repetition frequency (prf). Alternatively, for electrocardiographic studies it may be necessary to generate signals typical of normal and abnormal heart conditions, or $f(t)$ could be simply a train of pulses of carefully controlled shape to serve as a timing (or clock) signal. Regardless of the application, the frequency spectrum determines the frequencies required and the strengths of the in-phase and phase-quadrature components.

Unfortunately, almost any periodic function involves an infinity of harmonically related frequencies, and since it is not feasible to generate all of these, it is necessary to truncate the Fourier series. If the series converges rapidly, there may be only a handful of sinusoids whose amplitudes are significant, but for a piecewise-continuous function such as a square wave, the convergence is very slow indeed. To retain even a large number of terms now leads to an approximate $f(t)$ having noticeable defects, and some of these are illustrated in Fig. 4.9. If, for example, $f(t)$ is a timing signal, the critical parameter could be the rise time (inversely proportional to the maximum rate of change) or the time at which a particular level is achieved. For other applications it could be the maximum signal, in which case the overshoot is important, or even the ripple. Whichever is the case, the maximum error that can be tolerated in the critical parameter specifies the number of terms necessary for an adequate approximation to $f(t)$.

Steady-state response

In Section 3.10 we introduced the concept of the steady-state response of a stable system. This is the output that remains after all transients have decayed, and for the particular input $e^{i\omega t}$ with ω real, the steady-state response of a system whose transfer function is $H(s)$ is

$$x(t) = H(i\omega)e^{i\omega t}.$$

For the inputs $\cos \omega t$ and $\sin \omega t$ the corresponding outputs can be found by expressing the trigonometric functions in exponential form and using superposition, and thus if

$$\begin{aligned} f(t) &= a_1\cos \omega t + b_1 \sin \omega t \\ &= \tfrac{1}{2}(a_1 - ib_1)e^{i\omega t} + \tfrac{1}{2}(a_1 + ib_1)e^{-i\omega t}, \end{aligned} \tag{4.25}$$

the output is

$$x(t) = \tfrac{1}{2}(a_1 - ib_1)H(i\omega)e^{i\omega t} + \tfrac{1}{2}(a_1 + ib_1)H(-i\omega)e^{-i\omega t}. \tag{4.26}$$

If the input to the system is a periodic function $f(t)$, the Fourier-series representation provides a straightforward method for computing the steady-state response. From Eqs. (4.24) and (4.26) the output is

$$\begin{aligned} x(t) &= \frac{1}{2}a_0 H(0) \\ &\quad + \frac{1}{2}\sum_{n=1}^{\infty} \{(a_n - ib_n)H(in\omega)e^{in\omega t} + (a_n + ib_n)H(-in\omega)e^{-in\omega t}\}. \end{aligned} \tag{4.27}$$

For a real system whose transfer function is a real function of s,

$$H(-in\omega) = \overline{H(in\omega)}\ ,$$

where the bar denotes the complex conjugate, and Eq. (4.27) can be expressed in real form. Moreover, for a system of nonzero order,

$$H(s) \to 0 \quad \text{as} \quad |s| \to \infty$$

implying

$$|H(\pm in\omega)| \to 0 \quad \text{as} \quad n \to \infty.$$

The series representation of $x(t)$ therefore converges more rapidly than the series representation of the input $f(t)$, and this is consistent with the integrating or smoothing action of a system. Since there are few cases in which the series (4.27) can be summed analytically, the more rapid convergence is convenient for a numerical determination of $x(t)$.

Example. The periodic function $f(t) = t^2$ for $|t| < \pi$ is the input to a stable system whose transfer function can be approximated by

$$H(s) = \begin{cases} 1 - \left|\dfrac{s}{3}\right| & \text{for } |s| \leq 3, \\ 0 & \text{for } |s| > 3. \end{cases}$$

Find the steady-state response.

The period τ is 2π, implying $\omega = 1$, and since $f(t)$ is an even function,

$$b_n = 0$$

with (see Eq. (4.11))

$$a_n = \frac{2}{\pi}\int_0^{\pi} t^2 \cos nt\, dt.$$

Integration by parts gives

$$a_n = \frac{2}{\pi}\left[\frac{1}{n}\left(t^2 - \frac{2}{n^2}\right)\sin nt + \frac{2t}{n^2}\cos nt\right]_0^{\pi}$$

$$= \frac{4}{n^2}(-1)^n$$

for $n \neq 0$, with

$$a_0 = \frac{2\pi^2}{3}.$$

The Fourier-series representation of the periodic function $f(t)$ is therefore

$$f(t) = \frac{\pi^2}{3} + 4\sum_{n=1}^{\infty} \frac{(-1)^n}{n^2} \cos nt.$$

We remark in passing that since $f(t)$ is a continuous function with $f(0) = 0$, putting $t = 0$ in the series gives

$$\frac{1}{1^2} - \frac{1}{2^2} + \frac{1}{3^2} - \frac{1}{4^2} + \cdots = \frac{\pi^2}{12}.$$

Similarly, by putting $t = \pi$ and using the fact that $f(\pi) = \pi^2$, we obtain

$$\frac{1}{1^2} + \frac{1}{2^2} + \frac{1}{3^2} + \frac{1}{4^2} + \cdots = \frac{\pi^2}{6}.$$

These are two of the numerous series for integer powers of π^2 (through π^{26}) obtained by Leonhard Euler and published in 1748.

If $f(t)$ is the input to a system with transfer function $H(s)$, the steady-state response is

$$x(t) = \frac{\pi^2}{3}H(0) + 2\sum_{n=1}^{\infty} \frac{(-1)^n}{n^2}\{H(in)e^{int} + H(-in)e^{-int}\}.$$

With the given approximation to $H(s)$,

$$H(0) = 1, \qquad H(\pm i) = \tfrac{2}{3}, \qquad H(\pm 2i) = \tfrac{1}{3},$$

and

$$H(\pm in) = 0, \qquad \text{for } n \geq 3.$$

The output is therefore

$$x(t) = \tfrac{1}{3}(\pi^2 - 8\cos t + \cos 2t).$$

Exercises

Find the Fourier-series representation of

*1. $f(t) = \begin{cases} -\pi & \text{for } -\pi < t < 0, \\ t & \text{for } 0 < t < \pi. \end{cases}$

2. $f(t) = \begin{cases} t & \text{for } 0 < t < 1, \\ 3 - 2t & \text{for } 1 < t < \frac{3}{2}, \\ 0 & \text{for } \frac{3}{2} < t < 2. \end{cases}$

3. $f(t) = \sin t, \quad \text{for } 0 < t < \pi/2.$

4. $f(t) = e^{-}|t| \quad \text{for } |t| < 1.$

Find the Fourier-sine-series representation of

*5. $f(t) = \begin{cases} 1 - t & \text{for } 0 < t < 1, \\ 0 & \text{for } 1 < t < 2. \end{cases}$

6. $f(t) = \begin{cases} \sin t & \text{for } 0 < t < \pi, \\ 0 & \text{for } \pi < t < 2\pi. \end{cases}$

7. $f(t) = t^2, \quad \text{for } 0 < t < 1.$

8. $f(t) = \pi t - t^2, \quad \text{for } 0 < t < \pi.$

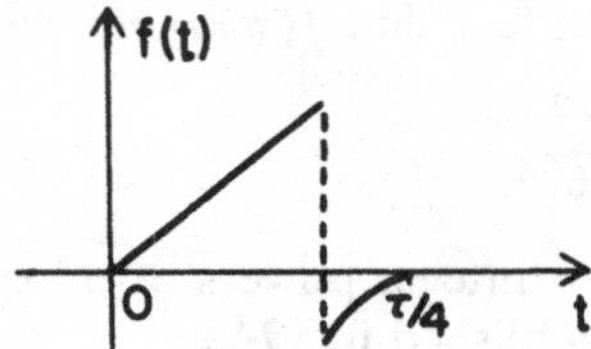

Fig. 4.11: First quarter-period of $f(t)$.

9. From the solution of the previous problem, deduce

$$\sum_{m=0}^{\infty} \frac{(-1)^m}{(2m+1)^3} = \frac{\pi^3}{32}.$$

Find the Fourier-cosine-series representation of

*10. $f(t) = \begin{cases} \sin t & \text{for } 0 < t < \pi, \\ 0 & \text{for } \pi < t < 2\pi. \end{cases}$

11. $f(t) = \begin{cases} 1 - t & \text{for } 0 < t < 1, \\ 0 & \text{for } 1 < t < 2. \end{cases}$

12. $f(t) = \begin{cases} 1 & \text{for } 0 < t < 1, \\ 2 - t & \text{for } 1 < t < 2. \end{cases}$

13. $f(t) = \begin{cases} 1 & \text{for } 0 < t < 1, \\ \frac{1}{2} & \text{for } 1 < t < 2, \\ 0 & \text{for } 2 < t < 4. \end{cases}$

14. Over its first quarter-period, $f(t)$ has the form shown in Fig. 4.11. Complete the waveform over the entire period $0 < t < \tau$ in the following cases:
(a) $f(t)$ is even and contains only even harmonics.
(b) $f(t)$ is even and contains only odd harmonics.
(c) $f(t)$ is even and contains both even and odd harmonics.
(d) $f(t)$ is odd and contains only even harmonics.
(e) $f(t)$ is odd and contains only odd harmonics.
(f) $f(t)$ is odd and contains both even and odd harmonics.

Each of the following periodic $f(t)$ is the input to a system whose transfer function is $H(s) = 1/(s+1)$. Find the resulting output and express it in real form.

15. $f(t) = |\cos 2t|$.

16. $f(t) = \left|\sin \dfrac{\pi t}{2l}\right|$, for $|t| < l$.

17. $f(t) = \sin \dfrac{\pi t}{2l}$, for $|t| < l$.

18. $f(t) = t^2$, for $|t| < 1$.

4.7 Complex form

When using the Fourier-series representation of a periodic function to find the steady-state response, we found it necessary to express the trigonometric functions in terms of exponentials, and this suggests the possibility of rewriting the series in exponential form to begin with.

From Eqs. (4.24) and (4.25),

$$f(t) = \frac{1}{2}a_0 + \sum_{n=1}^{\infty}\left\{\frac{1}{2}(a_n - ib_n)e^{in\omega t} + \frac{1}{2}(a_n + ib_n)e^{-in\omega t}\right\}$$

which can be written as

$$f(t) = \sum_{n=-\infty}^{\infty} c_n e^{in\omega t}, \tag{4.28}$$

where, for $n > 0$,

$$c_n = \frac{1}{2}(a_n - ib_n) = \frac{1}{T}\int_T f(t)(\cos n\omega t - i\sin n\omega t)\,dt$$

$$= \frac{1}{T}\int_T f(t)e^{-in\omega t}\,dt.$$

Similarly, for $n < 0$,

$$c_n = \frac{1}{2}(a_{-n} + ib_{-n}) = \frac{1}{T}\int_T f(t)(\cos n\omega t - i\sin n\omega t)\,dt$$

$$= \frac{1}{T}\int_T f(t)e^{-in\omega t}\,dt$$

and

$$c_0 = \frac{1}{2}a_0 = \frac{1}{T}\int_T f(t)\,dt,$$

showing that for all n

$$c_n = \frac{1}{\tau}\int_{\tau} f(t)e^{-in\omega t}\,dt, \tag{4.29}$$

where the integration is over any interval in t of length $\tau = 2\pi/\omega$.

The representation (4.28) with coefficients c_n given by Eq. (4.29) is called the *complex form* of the Fourier series. It is actually an exponential form – and this is a better description of (4.28) – but the word "complex" has now become standard. It could have been obtained directly without reference to the trigonometric series by using the fact that the exponentials $e^{in\omega t}$ with $n = 0, \pm 1, \pm 2, \ldots$ form a complete set of linearly independent functions, orthogonal over any $\tau = 2\pi/\omega$ range of t. The mathematical properties of the series are the same and, in particular, if $f(t)$ is piecewise-smooth, the series converges to

$$\tfrac{1}{2}\{f(t+) + f(t-)\}$$

at every point.

In contrast to the trigonometric form of the Fourier series there is now only one set of coefficients to be found, and integrals involving exponentials are easier to evaluate than those involving cosines and sines. The representation is also convenient for the determination of the steady-state response of a system, and if $H(s)$ is the transfer function, the steady-state response is simply

$$x(t) = \sum_{n=-\infty}^{\infty} c_n H(in\omega)e^{in\omega t}. \tag{4.30}$$

There are, however, disadvantages as well, and one of these is the extension of the frequency spectrum to negative frequencies. The spectrum now consists of the discrete frequencies $n\omega$ with $n = 0, \pm 1, \pm 2, \ldots$. The negative frequencies have no physical significance and they appear only as a result of the mathematical manipulation that converts cosines and sines into exponentials. The coefficients c_n representing the strengths of the frequency components are, in general, complex even if the function $f(t)$ is real. If $f(t)$ is real, implying that the coefficients a_n and b_n are real, it follows immediately from the way in which the c_n were defined that

$$c_{-n} = \bar{c}_n,$$

where the bar denotes the complex conjugate. This can also be obtained from Eq. (4.29) by using the fact that $\overline{f(t)} = f(t)$, and shows that the amplitudes of corresponding positive and negative frequencies are com-

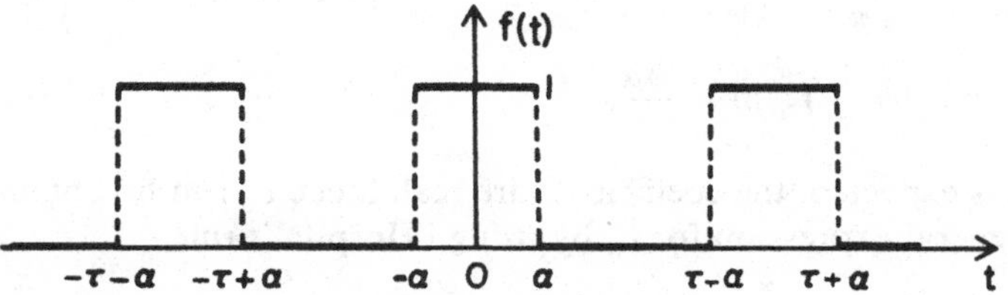

Fig. 4.12: The periodic function $f(t)$.

plex conjugates of one another. If, in addition, $f(t)$ is an even function,

$$c_n = c_{-n} = \bar{c}_n$$

and the coefficients are all real, whereas if $f(t)$ is odd,

$$c_n = -c_{-n} = -\bar{c}_n$$

and the coefficients are purely imaginary.

Finally, we note that the substitution of Eq. (4.29) into (4.28) gives

$$f(t) = \frac{1}{\tau} \sum_{n=-\infty}^{\infty} \int_{\tau} f(t') e^{in\omega(t-t')}\, dt', \tag{4.31}$$

where the left-hand side must be interpreted as the average of the left- and right-hand limits at any jump discontinuity in $f(t)$. This is the basis for the Fourier-integral representation discussed in Chapter 6.

Example. Find the complex Fourier-series representation of the periodic function $f(t)$ where

$$f(t) = \begin{cases} 1 & \text{for } 0 < t < \alpha, \\ 0 & \text{for } \alpha < t < \tau - \alpha, \\ 1 & \text{for } \tau - \alpha < t < \tau \end{cases}$$

with $\alpha < \tau/2$.

The periodic function $f(t)$ is shown in Fig. 4.12. Its period is clearly τ and if we choose the interval of integration to be $(-\alpha, \tau - \alpha)$, we have

$$\begin{aligned} c_n &= \frac{1}{\tau} \int_{-\alpha}^{\tau-\alpha} f(t) e^{-in\omega t}\, dt \\ &= \frac{1}{\tau} \int_{-\alpha}^{\alpha} e^{-in\omega t}\, dt \\ &= \frac{1}{\tau} \left[\frac{e^{-in\omega t}}{-in\omega} \right]_{-\alpha}^{\alpha} \qquad n \neq 0 \\ &= \frac{1}{n\pi} \sin n\omega\alpha, \end{aligned}$$

where $\omega = 2\pi/\tau$. Also

$$c_0 = \frac{1}{\tau}\int_{-\alpha}^{\alpha} dt = \frac{\omega\alpha}{\pi}$$

and, as expected, the coefficients are real. Since c_0 can be obtained from the general expression for c_n by using l'Hospital's rule,

$$f(t) = \sum_{n=-\infty}^{\infty} \frac{\sin n\omega\alpha}{n\pi} e^{in\omega t}.$$

Exercises

Find the complex Fourier-series representation of

*1. $f(t) = e^{-t}$, for $0 < t < 1$. 2. $f(t) = e^{-|t|}$, for $|t| < 1$.

3. $f(t) = t$, for $0 < t < \pi$. 4. $f(t) = |t|$, for $|t| < \pi$.

Suggested reading

Carslaw, H. S. *Introduction to the Theory of Fourier's Series and Integrals*. Cambridge University Press, Cambridge, UK, 1930. (Reprinted by Dover Publications, New York, 1946.) This is still the standard reference for Fourier series.

Churchill, R. V. *Fourier Series and Boundary Value Problems*. McGraw-Hill, New York, 1941. A good reference for the mathematical properties of Fourier series.

Kaplan, W. *Operational Methods for Linear Systems*. Addison-Wesley, Reading, MA, 1962.

Cheng, D. K. *Analysis of Linear Systems*. Addison-Wesley, Reading, MA, 1959. An elementary and practical presentation of Fourier series.

CHAPTER 5

Functions of a complex variable

In Chapter 1 we introduced some particular functions of a complex variable, such as powers and the exponential, that were needed for the chapters to follow. We now take up the subject again and develop a general theory of functions of a complex variable, including integration in the complex plane. Since the plane has no physical meaning it might seem that we are embarking on a study that has no relevance to an engineer, but a familiarity with complex-variable techniques is important in all branches of engineering. In particular, they are essential for a proper appreciation of the Fourier- and inverse Laplace-transform operations discussed in later chapters, and this is the reason for their consideration now.

5.1 General properties

We consider complex functions of the complex variable $z = x + iy$, where the definition of a function is analogous to that in real variable theory:

> If w and z are any two complex numbers, then w is a *function* of z (i.e., $w = f(z)$) if, to every value of z in some domain D, there corresponds one or more value(s) of w.

A *domain* is simply an open region of the complex plane bounded by a closed contour, and we shall use such terms with no more precise definition than geometrical intuition requires.

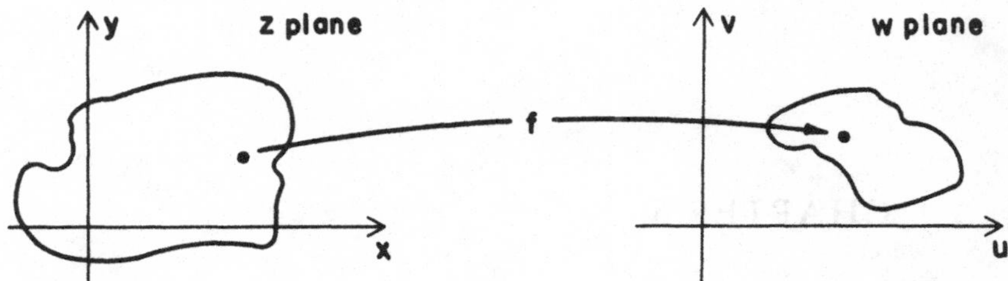

Fig. 5.1: The mapping provided by a function f.

Since w is a complex number, it can be separated into real and imaginary parts, and we shall write

$$w = u + iv,$$

where

$$u = u(x, y) = \operatorname{Re} w,$$
$$v = v(x, y) = \operatorname{Im} w$$

are real functions of the two real variables x and y. If $f(z)$ is a single-valued function of z (e.g., $f(z) = z^2$), then u and v are single-valued functions of x and y, whereas for a multivalued function (e.g., $f(z) = z^{1/2}$), u and v are also multivalued. For the moment at least we confine our attention to functions that are single-valued, so that a knowledge of the function uniquely determines w and, hence, u and v. In any given instance it is a trivial matter to find u and v. Thus, if $f(z) = z^2$, we have

$$\begin{aligned} w &= (x + iy)^2 \\ &= x^2 - y^2 + 2ixy. \end{aligned}$$

Separation into real and imaginary parts then gives

$$u(x, y) = x^2 - y^2, \qquad v(x, y) = 2xy. \tag{5.1}$$

Mapping

A function f provides a mapping of the complex z plane onto a complex w plane as shown in Fig. 5.1, and the transformation from one plane to another is often used in potential theory and other disciplines to simplify the solution of boundary-value problems. It is also of interest in systems theory.

Example 1. Consider $f(z) = 1/(a + z)$ for real $a > 0$.

Since

$$\frac{1}{a+z} = \frac{1}{a+x+iy} = \frac{a+x-iy}{(a+x)^2+y^2},$$

it follows that

$$u(x,y) = \frac{a+x}{(a+x)^2+y^2}, \qquad v(x,y) = -\frac{y}{(a+x)^2+y^2}.$$

For every point x, y in the z plane there is a corresponding "image" point u, v in the w plane that is found by assigning the appropriate values to x and y. A point on the positive real axis of the z plane has $y = 0$ with $x \geq 0$, and since

$$u(x,0) = \frac{1}{a+x}, \qquad v(x,0) = 0,$$

this is mapped into a point on the positive real axis of the w plane. Indeed, as x varies from 0 to ∞, we trace out the real axis of the w plane from $1/a$ back to 0. Similarly, for a point on the imaginary axis of the z plane,

$$u(0,y) = \frac{a}{a^2+y^2}, \qquad v(0,y) = -\frac{y}{a^2+y^2}$$

and by eliminating y from these "parametric" equations we obtain

$$\left(u - \frac{1}{2a}\right)^2 + v^2 = \left(\frac{1}{2a}\right)^2,$$

representing a circle of radius $1/(2a)$ centered on the point $u = 1/(2a)$, $v = 0$. As y varies from $-\infty$ through 0 to ∞ the corresponding point in the w plane describes the circle moving clockwise from the origin. The entire fourth and first quadrants of the z plane are mapped onto the upper and lower semicircular regions, respectively, and in particular, the positive imaginary axis is mapped onto the lower perimeter shown in Fig. 5.2.

The mapping of the positive imaginary axis is employed in control theory to determine the stability of a system. For a simple *R-L* circuit the transfer function is proportional to

$$H(s) = \frac{1}{a+s}$$

with $a = R/L$, and this is the same as our function $f(z)$ with z replaced by s. The lower semicircle in Fig. 5.2 is the plot of $H(i\omega)$ versus ω, $0 \leq \omega < \infty$, and is referred to as a *Nyquist diagram*, after the U.S. engineer Harry Nyquist (1889–1976).

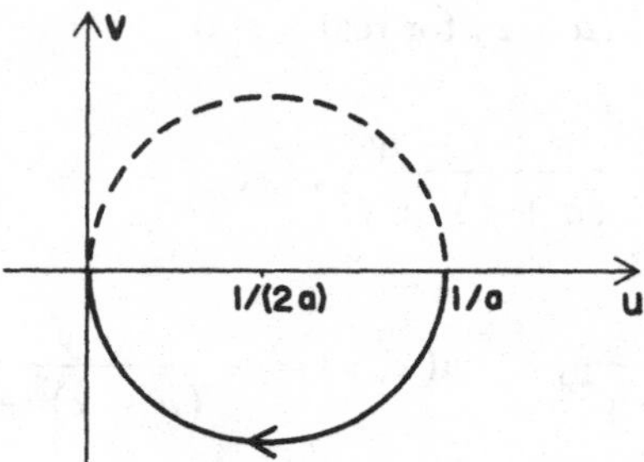

Fig. 5.2: The lower semicircle is traced out as z varies from 0 to $i\infty$.

Example 2. Examine the mapping provided by the function $f(z) = z^2$.

The functions $u(x, y)$ and $v(x, y)$ are given in Eq. (5.1), and by eliminating x from these relations, we obtain

$$u = \left(\frac{v}{2y}\right)^2 - y^2,$$

that is,

$$v^2 = 4y^2(u + y^2),$$

representing a parabola of *latus rectum* $4y^2$ with focus at the origin. Similarly, by eliminating y from the expressions for u and v,

$$u = x^2 - \left(\frac{v}{2x}\right)^2,$$

which can be rearranged as

$$v^2 = 4x^2(x^2 - u).$$

This is a parabola of *latus rectum* $4x^2$ with focus at the origin, but "pointed" in the opposite direction to the previous one. Thus, the grid lines $y =$ constant and $x =$ constant are mapped into intersecting parabolas as shown in Fig. 5.3.

The half-plane $y > 0$ maps into the entire w plane except for the positive u axis, which is the image of the x axis, and the half-plane $y < 0$ has a similar map. The grid lines $y = \pm c_1$ are both mapped onto the same parabola, as are the lines $x = \pm c_2$, and the image of the point x, y is, in fact, the same as that of $-x, -y$. There is nothing wrong with this because f is still single-valued, but it is obvious that difficulties would arise if we tried to carry out the reverse transformation from the w plane back to the z plane. The transformation to do this is $z = w^{1/2}$, which is a multivalued function of w.

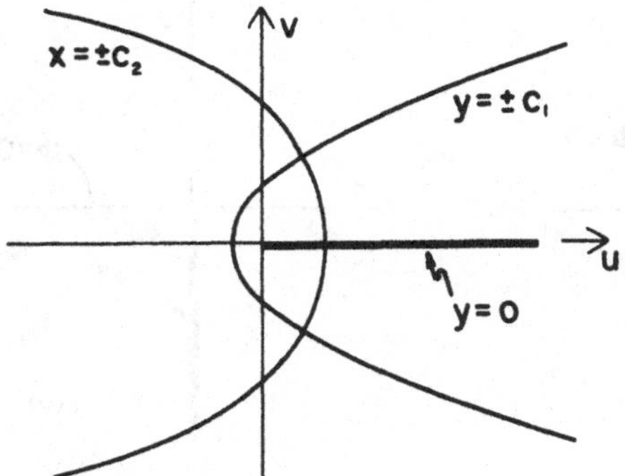

Fig. 5.3: The mapping $f(z) = z^2$.

Multivalued functions

In Section 1.4 we saw that any fractional power of z, like $z^{1/n}$ where n is an integer, is a multivalued function of z, and the mere statement that $f(z) = z^{1/n}$ does not uniquely determine the function. In order to operate with functions of this type and to develop a general theory that will include them, it is necessary to find some way to pin down the function.

The multivaluedness of a function is attributable to the argument of a complex number. When the complex number z is written in the polar form $re^{i\theta}$ we have

$$r = |z| \quad \text{and} \quad \theta = \arg z,$$

and if θ_1 is one possible value of $\arg z$, the full range of values of θ is

$$\theta = \theta_1 + 2m\pi, \tag{5.2}$$

where m is any integer (positive, negative, or zero). The indefinite nature of θ does *not* make z multivalued (since $\cos\theta = \cos\theta_1$ and $\sin\theta = \sin\theta_1$, the real and imaginary parts of z are independent of m), but it *does* make functions such as fractional powers of z and $\log z$ multivalued.

Consider, for example, $z^{1/2}$. By inserting the polar form for z we have

$$z^{1/2} = |z|^{1/2}e^{i\phi},$$

where

$$\phi = \arg z^{1/2} = \frac{\theta}{2}\left(= \frac{\theta_1}{2} + m\pi\right). \tag{5.3}$$

Let us examine the behavior of the function as we move around a circle of unit radius centered on the origin. We start at $z = 1$, where we presume the function to be known. For simplicity, we choose $\theta = 0$

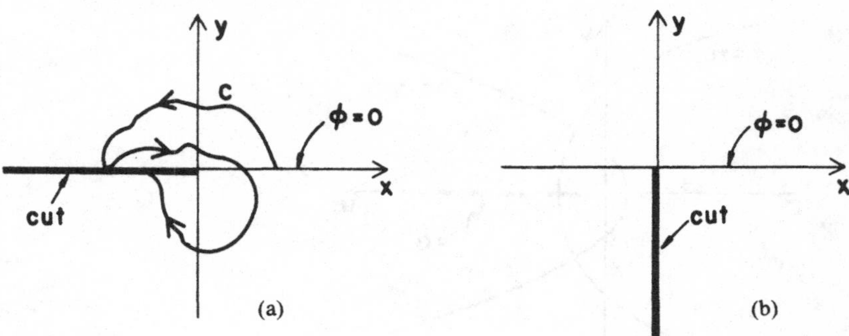

Fig. 5.4: Two examples of the specification of $z^{1/2}$ in cut planes.

there, implying $\phi = 0$, and since $|z|^{1/2} = 1$, it follows that $z^{1/2} = 1$. At $z = i$, $z^{1/2} = e^{i\pi/4} = (1 + i)/\sqrt{2}$, and so on around the circle, with θ and, hence, ϕ increasing continuously. When we get back to $z = 1$, θ has increased to 2π and $z^{1/2}$ has become $e^{i\pi} = -1$. We have now picked up the function having $m = 1$ in Eq. (5.3). If we continue around the circle with θ still increasing, we arrive back at $z = 1$ with $\theta = 4\pi$, and although this is equivalent to the choice $m = 2$ in Eq. (5.3), the key point is we have recovered the value of $z^{1/2}$ which we started with. By circling the origin repeatedly, we keep on cycling between the two sets of function values. We therefore say that $z^{1/2}$ is a *double- (or two-) valued function*, and this is true for all z except $z = 0$, where the function is zero. The two sets of values are referred to as *branches* of the function.

If we start with a particular choice of $z^{1/2}$ at (let us say) $z = 1$ and circle around in a way that does not include the origin, we come back to the original value. Try, for example, going from $z = 1$ to $z = 2 + i$, and thence to $1 + 2i$, i, and back to 1. Clearly, the troublesome point is the origin, and in order to define the function uniquely, it is necessary to prevent the origin from being circled. We can do this by limiting the domain such that ϕ is made single-valued; we alluded to this approach in Section 1.4.

Suppose that we place a barrier or cut in the z plane extending outward from $z = 0$. The cut must extend to infinity to eliminate the possibility of sneaking around the far end, and we can imagine making such a cut with a pair of scissors. We do not have to be particularly careful about the cut as long as it goes from the origin to infinity, but since we will ultimately have to show or describe the cut that we have made, it may be convenient to have it straight. As an example, let us cut along the negative real axis as shown in Fig. 5.4(a) and, in accordance with Eq. (5.3), choose ϕ to be zero on the positive real axis. As the point

z moves away from the positive real axis, ϕ changes continuously, and by following a path C like that in Fig. 5.4(a) we can reach the upper side of the cut where $\phi = \pi/2$. This result is independent of the path, provided we remain in the domain. In particular, the cut is a barrier that cannot be crossed, and to find ϕ on the lower side we can imagine having to wend our way back around the origin with ϕ decreasing continuously until we finally reach the lower side where $\phi = -\pi/2$. Thus, $\phi = \arg z^{1/2}$ is discontinuous across the cut, and hence $z^{1/2}$ is discontinuous. This is hardly surprising: If it were not discontinuous, there would be no need for the cut at all. The important point is that in the *cut plane* with $z^{1/2}$ specified at a single location (it is sufficient to know $\arg z^{1/2}$ there), the function $f(z) = z^{1/2}$ is *uniquely defined* everywhere, and is therefore a single-valued function of z.

Different cuts combined with different specifications of $\arg z^{1/2}$ at a particular point define different functions. We call these functions *branches* of $z^{1/2}$ and the cuts *branch cuts*. The origin whose encirclement produced the double-valuedness of $z^{1/2}$ is known as the *branch point*. There is no unique way of choosing the cut or of dividing the function into its branches, but since the function is discontinuous across a cut, it may be convenient to place it in the portion of the z plane of least interest. If, for example, we are mainly concerned with values of z for which $\operatorname{Im} z \geq 0$, the branch cut could be taken as shown in Fig. 5.4(b). In the first, second, and fourth quadrants the functions defined in Fig. 5.4(a) and (b) are the same, but since they differ in the third quadrant, the functions are not the same overall. Nevertheless, each is a single-valued function representing $z^{1/2}$ in the cut plane.

For the more general function $z^{1/n}$ where n (> 1) is an integer, the branch point is again $z = 0$ and the z plane must be cut in the manner described above. Without such a cut and the specification of $z^{1/n}$ at a particular point, the function is not unique. Likewise, $(z+1)^{1/n}$ has a branch point at $z = -1$ and $(z^2 + a^2)^{1/n}$ has branch points at $z = \pm ia$. Another example of a multivalued function is $\log z$. From Eq. (1.24),

$$\log z = \ln z + i \arg z$$

and since

$$\arg z = \theta_1 + 2m\pi$$

(see Eq. (5.2)), every integer value of m gives a branch. Therefore, $\log z$ is infinitely many-valued. The branch point is again $z = 0$ and we can eliminate the multiplicity by introducing a cut extending outward from $z = 0$. If the cut is along the negative real axis with (let us say) $\log 1 = 0$, $\arg z$ is confined to the range $(-\pi, \pi)$ and the resulting branch is the

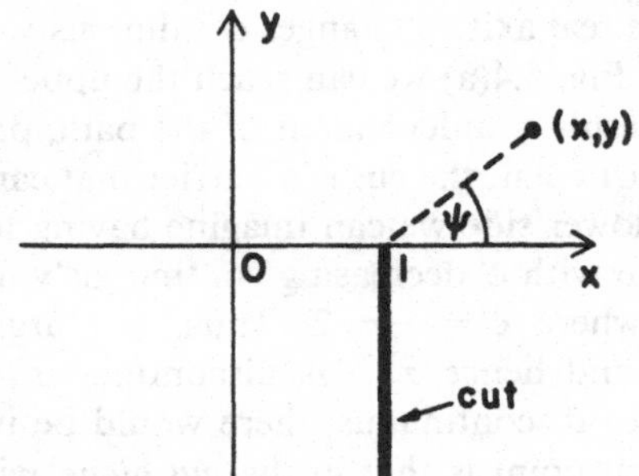

Fig. 5.5: Cut plane used for the determination of $(z-1)^{1/3}$.

principal value of $\log z$ defined in Section 1.4. The treatment of the general power ξ^z is similar.

From now on it will be assumed that the functions we are dealing with are single-valued in some domain. In the case of functions that are potentially multivalued, the responsibility for specifying the branch cuts and the value at a given point, both of which are necessary for a unique definition, rests with the person who originates the function. This is important in any calculation. Those who follow you should not have to guess the function you really had in mind.

Example. If $f(z)=(z-1)^{1/3}$ in the cut plane shown in Fig. 5.5 with $f(0)=-1$, determine $f(1+i)$ and $f(2)$.

It is convenient to introduce the polar angle ψ shown in the figure. Then $-\pi/2<\psi<3\pi/2$ with

$$\cos\psi=\frac{x-1}{\left\{(x-1)^2+y^2\right\}^{1/2}},\qquad \sin\psi=\frac{y}{\left\{(x-1)^2+y^2\right\}^{1/2}},$$

and

$$f(z)=|z-1|^{1/3}e^{i\phi},$$

where

$$\phi=\arg(z-1)^{1/3}=\tfrac{1}{3}(\psi+2m\pi)$$

for one of the choices $m=0$, 1, or 2.

The three choices correspond to the three possible branches of $(z-1)^{1/3}$, and the desired branch is obtained by enforcing the condition $f(0)=-1$. At the origin, $x=y=0$ and $\psi=\pi$. Also $|z-1|^{1/3}=|-1|^{1/3}=1$ and hence

$$f(0)=e^{i(1+2m)\pi/3},$$

which equals -1 if $m = 1$. The required function is therefore

$$f(z) = |z - 1|^{1/3} e^{i(\psi + 2\pi)/3},$$

where ψ is defined above. When $z = 1 + i$ we have $|z - 1|^{1/3} = 1$ and $\psi = \pi/2$, giving

$$f(1 + i) = e^{5i\pi/6} = \tfrac{1}{2}(-\sqrt{3} + i),$$

and when $z = 2$, $|z - 1|^{1/3} = 1$ and $\psi = 0$, giving

$$f(2) = e^{2i\pi/3} = \tfrac{1}{2}(-1 + i\sqrt{3}).$$

Exercises

1. Under the mapping $w = (1 + i)z - 2$, find the image of:
   *(i) The lines $x = 0$ and $y = 0$.
   (ii) The region $|z - 2| \leq 2$.

2. Under the mapping $w = 1/[z(z + 1)]$, find the image of the lines:
   (i) $z = iy$, $0 < y < \infty$. (ii) $z = -1 + iy$, $0 < y < \infty$.

3. Find the images of the following curves under the mapping $w = z^2$.
   (i) $x = 1$. (ii) $y = x$.
   (iii) $y = 1 - x$. (iv) $y^2 = x^2 + 1$.

4. Find the images of the following circles and lines under the mapping $w = 1/z$.
   (i) $|z| = 1$. (ii) $|z + 1| = 1$.
   (iii) $y = x + 1$. (iv) $x = 1$.

5. Find the image of the region $1 < x < 2$ under the mapping $w = 1/z$.

6. Prove that $w = (1 + iz)/(i + z)$ maps the part of the real axis between $z = -1$ and $z = 1$ onto a semicircle in the w plane.

7. Identify the branch points, if any, of the following functions.
   (i) $(z^2 - 2z)^{-1/2}$. (ii) $(z + 1)^{1/2} + (z - 1)^{1/2}$.
   (iii) $(z^{1/2})^2$. (iv) z^z.

*8. If $f(z) = z^{1/2}$ in a plane cut along the positive imaginary axis with $f(1) = 1$, determine $f(-i)$ and $f(-1)$.

9. If $f(z) = (z^2 - 1)^{1/2}$ in a plane cut from $z = -1$ to $z = -1 - i\infty$ and from $z = 1$ to $z = 1 - i\infty$, and if $f(0) = -i$, determine $f(2)$, $f(2i)$, $f(-2)$, and $f(-2i)$. [*Hint*: Write $f(z) = (z + 1)^{1/2}(z - 1)^{1/2}$ and introduce a polar angle ψ associated with each branch point.]

10. (i) Repeat the previous exercise using a plane with a *single* cut from $z = -1$ to $z = 1$ that passes just below the origin.
(ii) Why does a single cut suffice to make $(z^2 - 1)^{1/2}$ single-valued, and would it be adequate for $(z^2 - 1)^{1/3}$?

11. If $w = \arcsin z$, then

$$z = \sin w = \frac{1}{2i}(e^{iw} - e^{-iw}),$$

giving

$$e^{2iw} - 2ize^{iw} - 1 = 0.$$

By solving this quadratic equation, we obtain

$$e^{iw} = iz \pm \sqrt{1 - z^2}$$

and therefore

$$w = \arcsin z = -i\log(iz + \sqrt{1 - z^2}),$$

where we have replaced the $\pm$ with $+$ for simplicity, since we will need to select a branch for $\sqrt{1 - z^2}$ anyway. In this manner, show:

(i) $\arccos z = -i\log(z + \sqrt{z^2 - 1})$.
(ii) $\arctan z = \dfrac{i}{2}\log\dfrac{1 - iz}{1 + iz}$.
(iii) $\operatorname{arcsinh} z = \log(z + \sqrt{z^2 + 1})$.
(iv) $\operatorname{arccosh} z = \log(z + \sqrt{z^2 - 1})$.
(v) $\operatorname{arctanh} z = \dfrac{1}{2}\log\dfrac{1 + z}{1 - z}$.

12. Show that $\arccos z = -i\log(z + \sqrt{z^2 - 1})$ is infinitely many-valued, and define branch cuts sufficient to make it single-valued.

Continuity and differentiability

Even if we confine attention to functions $f(z)$ that are single-valued in some domain D, as we shall do from now on, the definition of a function, which was given at the beginning of this section, is too broad to be of much interest. According to the definition, a function of a complex variable z is exactly the same as a complex function

$$u(x, y) + iv(x, y)$$

of the two real variables x and y, and it is only by restricting the types of functions considered that the unique properties of functions of a complex variable are revealed. We shall do so in two steps, starting with the requirement of continuity.

The definition of continuity is the same as that for a function of a real variable and accords with our physical intuition: If $f(z)$ is continuous at z_0, we can make $f(z)$ as close as we desire to the value $f(z_0)$ by taking z sufficiently close to z_0. Mathematically expressed,

> $f(z)$ is *continuous* at the point z_0 in D if $f(z_0)$ exists and if, for any $\varepsilon > 0$, there exists δ such that
>
> $|f(z) - f(z_0)| < \varepsilon$
>
> for all z in D satisfying $|z - z_0| < \delta$.

The function is continuous in the domain D if it is continuous at every point z_0 in D. In general δ will depend on z_0 as well as ε, but if it is possible to find $\delta = \delta(\varepsilon)$ independent of z_0 such that $|f(z) - f(z_0)| < \varepsilon$ for all z, z_0 in D satisfying $|z - z_0| < \delta$, $f(z)$ is said to be *uniformly continuous* in D. A function that is continuous in a bounded closed domain is uniformly continuous there. The set of points z such that $|z - z_0| < \delta$ for some $\delta > 0$ is called a *neighborhood* of z_0, and is simply the interior of a circle of radius δ about the point in question.

It is easy to show that if $f(z)$ is continuous, $u(x, y)$ and $v(x, y)$ are continuous functions of x and y. Conversely, if u and v are continuous, then $f(z)$ is a continuous function of z, and continuous functions of continuous functions are likewise continuous.

The next requirement that we impose is that of differentiability, and the only functions $f(z)$ of practical concern are those to which the process of differentiation can be applied. By definition

> $f(z)$ is *differentiable* at the point z_0 in D if
>
> $$\frac{f(z) - f(z_0)}{z - z_0}$$
>
> tends to a unique limit as $z \to z_0$.

If the limit exists it is called the *derivative* of $f(z)$ at $z = z_0$ and is denoted by $f'(z_0)$. The definition asserts that for a given $\varepsilon > 0$ there exists δ such that

$$\left| \frac{f(z) - f(z_0)}{z - z_0} - f'(z_0) \right| < \varepsilon$$

for all z in D satisfying $|z - z_0| < \delta$.

In appearance the definition is the same as that for a function of a real variable, and because of this the familiar results hold (e.g., $(d/dz)z^n = nz^{n-1}$) as do the real variable rules for differentiating sums, products, and quotients of functions. Nevertheless, the definition has some remarkable consequences that have no counterpart in real variable theory, and these follow from the seemingly innocuous requirement that the limit must be *unique* regardless of the manner in which z is allowed to approach z_0. If t is a real variable, $\Delta t = t - t_0$ can tend to zero only through positive and negative values; we have previously met functions whose right- and left-hand derivatives exist but differ at a point (e.g., $tu(t)$ at $t = 0$ where $u(t)$ is the unit-step function). In complex-variable theory, however, the limit involved in the definition of $f'(z)$ is two-dimensional. There are an infinity of ways in which $\Delta z = z - z_0$ can tend to zero, and no meaning is given to a directional derivative as such. The derivative exists if and only if the limit is the same whether z "barrels into" z_0 in some fixed direction or, for example, spirals around in an arbitrary manner.

Differentiability is a more stringent requirement than continuity. If the derivative $f'(z_0)$ exists, then

$$\lim_{z\to z_0}\{f(z)-f(z_0)\} = \lim_{z\to z_0}\left\{\frac{f(z)-f(z_0)}{z-z_0}\right\}\lim_{z\to z_0}(z-z_0)$$

$$= f'(z_0)\lim_{z\to z_0}(z-z_0) = 0,$$

that is,

$$\lim_{z\to z_0} f(z) = f(z_0),$$

which shows that $f(z)$ is continuous at z_0. Thus, differentiability implies continuity, but continuity does *not* imply differentiability, as evidenced by the function $f(z) = |z|^2$. Since $u(x, y) = x^2 + y^2$ and $v(x, y) = 0$, u and v are continuous functions of x and y, and $f(z)$ is therefore continuous at every point. However, $f(z)$ is differentiable only at the origin. To see this, consider

$$\frac{f(z)-f(z_0)}{z-z_0} = \frac{|z|^2-|z_0|^2}{z-z_0} = \frac{z\bar{z}-z_0\bar{z}_0}{z-z_0},$$

which can be rewritten as

$$\frac{f(z)-f(z_0)}{z-z_0} = \bar{z} + z_0\frac{\bar{z}-\bar{z}_0}{z-z_0}$$

$$= \bar{z} + z_0(\cos 2\phi - i\sin 2\phi),$$

where $\phi = \arg(z - z_0)$. As $z \to z_0$, $\bar{z} \to \bar{z}_0$ but the limit of the second term depends on ϕ, that is, on the direction in which z approaches z_0. It follows that the limit is not unique unless $z_0 = 0$, in which case the second term is absent.

5.2 Analytic functions

A function $w = f(z)$ that is differentiable at every point of a domain D is said to be an *analytic* function in D. The terms *regular* and *holomorphic* are also used and are synonymous with analytic.

Almost the entire theory of functions of a complex variable is confined to a study of these functions, and the theory is indelibly associated with the work of one man: Augustin-Louis Cauchy (1789–1857), the French engineer turned mathematician. He sought to give to the methods of analysis the same rigor that was then demanded in geometry. His 1814 memoir on definite integrals with complex limits inaugurated his career as the developer of the theory of functions of a complex variable, and with the publication of his 1847 paper on the imaginaries, the theory as we now know it was virtually complete. This is not, of course, to deny the contributions of others. The German mathematicians Karl Weierstrass (1815–97) and Georg Friedrich Bernhard Riemann (1826–66) independently pursued the development of Cauchy's theory along other distinct lines. When Cauchy began publishing his brilliant discoveries, it is said that Karl Friedrich Gauss (1777–1855), the German mathematician who has been called the Prince of Mathematics, ignored them. The fact of the matter is that he had obtained the same results earlier and had described them in a letter to Friedrich Wilhelm Bessel in 1811, but had not felt them worthy of publication. Fortunately, this did not deter Cauchy, and his pursuit of what would now be regarded as a topic in pure mathematics has provided us with tools of considerable importance for the solution of physical problems.

Cauchy–Riemann equations

The necessary and sufficient conditions for a function to be analytic are contained in the following theorem.

Theorem: If $w = u + iv = f(z)$ is analytic in D, then the first partial derivatives of u and v exist in D and satisfy the Cauchy–Riemann equations

$$\frac{\partial u}{\partial x} = \frac{\partial v}{\partial y}, \qquad \frac{\partial u}{\partial y} = -\frac{\partial v}{\partial x} \tag{5.4}$$

in D. Furthermore

$$f'(z) = \frac{\partial u}{\partial x} + i\frac{\partial v}{\partial x} = \frac{\partial v}{\partial y} + i\frac{\partial v}{\partial x} = \frac{\partial u}{\partial x} - i\frac{\partial u}{\partial y} = \frac{\partial v}{\partial y} - i\frac{\partial u}{\partial y}. \tag{5.5}$$

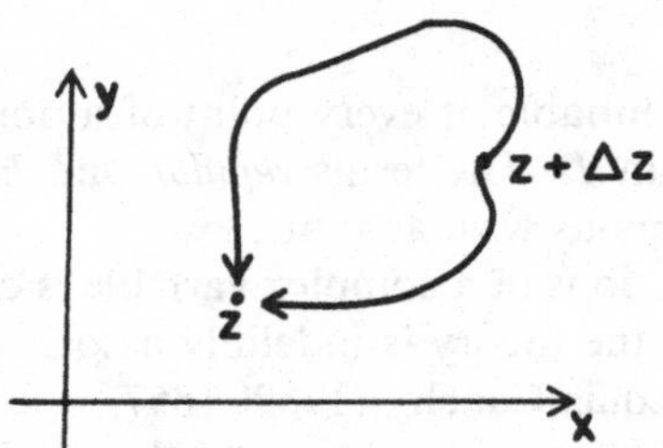

Fig. 5.6: Paths used in the theorem.

Conversely, if u and v have continuous first derivatives in D and satisfy the Cauchy–Riemann equations, then $w = u + iv = f(z)$ is analytic in D.

The first part of the theorem gives necessary conditions for analyticity, and the second part states that if the partial derivatives are continuous as well, the conditions are sufficient to ensure analyticity.

Proof: Let z be a point of D. If $f(z) = u(x, y) + iv(x, y)$ is analytic in D, the derivative

$$f'(z) = \lim_{\Delta z \to 0} \frac{f(z + \Delta z) - f(z)}{\Delta z}$$

exists and is independent of the manner in which Δz approaches zero. In general, $\Delta z = \Delta x + i\,\Delta y$, but we shall now choose two particular paths along which $\Delta z \to 0$ (see Fig. 5.6) and require that the results be the same. For the first of these Δz is wholly real so that $\Delta y = 0$, and thus

$$\begin{aligned} f'(z) &= \lim_{\Delta x \to 0} \frac{u(x + \Delta x, y) + iv(x + \Delta x, y) - u(x, y) - iv(x, y)}{\Delta x} \\ &= \lim_{\Delta x \to 0} \frac{u(x + \Delta x, y) - u(x, y)}{\Delta x} \\ &\quad + i \lim_{\Delta x \to 0} \frac{v(x + \Delta x, y) - v(x, y)}{\Delta x} \\ &= \frac{\partial u}{\partial x} + i\frac{\partial v}{\partial x} \end{aligned} \tag{5.6}$$

by the definition of partial derivative. Along the second path Δz is purely imaginary so that $\Delta x = 0$, and then

$$f'(z) = \lim_{\Delta y \to 0} \frac{u(x, y + \Delta y) + iv(x, y + \Delta y) - u(x, y) - iv(x, y)}{i \Delta y}$$

$$= -i \lim_{\Delta y \to 0} \frac{u(x, y + \Delta y) - u(x, y)}{\Delta y}$$

$$+ \lim_{\Delta y \to 0} \frac{v(x, y + \Delta y) - v(x, y)}{\Delta y}$$

$$= -i \frac{\partial u}{\partial y} + \frac{\partial v}{\partial y}. \tag{5.7}$$

The existence of $f'(z)$ therefore implies that the four partial derivatives in Eqs. (5.6) and (5.7) exist, and since the derivative is unique, we can calculate it using either (5.6) or (5.7). By equating the real and imaginary parts we obtain the Cauchy–Riemann equations (5.4), from which (5.5) follows immediately.

We now turn to the second (converse) part of the theorem for which it is no longer enough to choose only a particular path. Since u and its first partial derivatives are continuous, the chain rule for differentials can be employed to give

$$\Delta u = u(x + \Delta x, y + \Delta y) - u(x, y)$$

$$= \frac{\partial u}{\partial x} \Delta x + \frac{\partial u}{\partial y} \Delta y + \varepsilon_1 \Delta x + \varepsilon_2 \Delta y,$$

where $\varepsilon_1, \varepsilon_2 \to 0$ as Δx and $\Delta y \to 0$. There is a similar formula for Δv, and hence

$$\Delta f = f(z + \Delta z) - f(z) = \Delta u + i \Delta v$$

$$= \frac{\partial u}{\partial x} \Delta x + \frac{\partial u}{\partial y} \Delta y + \varepsilon_1 \Delta x + \varepsilon_2 \Delta y$$

$$+ i \left(\frac{\partial v}{\partial x} \Delta x + \frac{\partial v}{\partial y} \Delta y + \varepsilon_3 \Delta x + \varepsilon_4 \Delta y \right).$$

If the Cauchy–Riemann equations are satisfied, we can replace $\partial u/\partial y$ by $-\partial v/\partial x$ and $\partial v/\partial y$ by $\partial u/\partial x$ and write the last equation as

$$\Delta f = \left(\frac{\partial u}{\partial x} + i \frac{\partial v}{\partial x} \right) \Delta z + \delta_1 \Delta x + \delta_2 \Delta y,$$

where $\Delta z = \Delta x + i \Delta y$ and δ_1 $(= \varepsilon_1 + i\varepsilon_3), \delta_2 (= \varepsilon_2 + i\varepsilon_4) \to 0$ as $\Delta z \to 0$. Thus

$$\frac{\Delta f}{\Delta z} = \frac{\partial u}{\partial x} + i \frac{\partial v}{\partial x} + \delta_1 \frac{\Delta x}{\Delta z} + \delta_2 \frac{\Delta y}{\Delta z}$$

and since

$$\left|\frac{\Delta x}{\Delta z}\right|, \left|\frac{\Delta y}{\Delta z}\right| \leq 1,$$

the terms involving δ_1 and δ_2 tend to zero with Δz. It follows that

$$f'(z) = \lim_{\Delta z \to 0} \frac{\Delta f}{\Delta z} = \frac{\partial u}{\partial x} + i\frac{\partial v}{\partial x}$$

showing that the derivative $f'(z)$ exists. This completes the proof. □

The theorem provides a perfect test for analyticity: If $f(z)$ is analytic, the partial derivatives exist and satisfy the Cauchy–Riemann equations, and if the derivatives are continuous and satisfy the equations, then $f(z)$ is analytic where this occurs.

Example 1. Determine the analyticity of the function $w = f(z) = z^3$.
Since $z^3 = (x + iy)^3 = x^3 + 3ix^2y - 3xy^2 - iy^3$ we have

$$u(x, y) = x^3 - 3xy^2, \qquad v(x, y) = 3x^2y - y^3$$

and

$$\frac{\partial u}{\partial x} = 3x^2 - 3y^2, \qquad \frac{\partial u}{\partial y} = -6xy,$$

$$\frac{\partial v}{\partial x} = 6xy, \qquad \frac{\partial v}{\partial y} = 3x^2 - 3y^2.$$

All four partial derivatives exist and are continuous for all finite x, y, and the Cauchy–Riemann equations are satisfied. Hence, $w = z^3$ is analytic for all finite z, that is, in the finite part of the complex plane.

Example 2. Determine the analyticity of the function $w = f(z) = 1/z$.
Since

$$\frac{1}{z} = \frac{\bar{z}}{z\bar{z}} = \frac{x - iy}{x^2 + y^2}$$

we have

$$u(x, y) = \frac{x}{x^2 + y^2}, \qquad v(x, y) = -\frac{y}{x^2 + y^2}$$

and

$$\frac{\partial u}{\partial x} = \frac{y^2 - x^2}{(x^2 + y^2)^2} = \frac{\partial v}{\partial y}, \qquad \frac{\partial u}{\partial y} = -\frac{2xy}{(x^2 + y^2)^2} = -\frac{\partial v}{\partial x}.$$

The Cauchy–Riemann equations are satisfied and the partial derivatives exist and are continuous except when $x^2 + y^2 = 0$ (i.e., $z = 0$). Hence, $w = 1/z$ is analytic except at $z = 0$.

This illustrates the fact that a function can be analytic except at isolated points which are called *singularities* (or singular points) of the function. Since analyticity implies a domain, we do not apply the term analytic to functions whose derivatives exist only at certain points.

Example 3. Determine the analyticity of the function $w = f(z) = z\bar{z}$.

Since $z\bar{z} = x^2 + y^2$ we have

$$u(x, y) = x^2 + y^2, \qquad v(x, y) = 0$$

and

$$\frac{\partial u}{\partial x} = 2x, \qquad \frac{\partial u}{\partial y} = 2y$$

with $\partial v/\partial x = \partial v/\partial y = 0$. The Cauchy–Riemann equations are satisfied only if $x = y = 0$ (i.e., at $z = 0$). Hence, $w = z\bar{z}$ is nowhere analytic.

In this third example there is a feature of $f(z)$ that shows immediately that it cannot be analytic. Consider the function $w = u + iv$ where $u = u(x, y)$ and $v = v(x, y)$. Since

$$x = \frac{1}{2}(z + \bar{z}), \qquad y = \frac{1}{2i}(z - \bar{z}), \tag{5.8}$$

u and v can be regarded as functions of z and $\bar{z}$. If u and v have continuous first derivatives with respect to x and y, the application of the chain rule for differentiation gives

$$\begin{aligned}\frac{\partial w}{\partial \bar{z}} &= \frac{\partial}{\partial \bar{z}}(u + iv)\\ &= \frac{\partial u}{\partial x}\frac{\partial x}{\partial \bar{z}} + \frac{\partial u}{\partial y}\frac{\partial y}{\partial \bar{z}} + i\left(\frac{\partial v}{\partial x}\frac{\partial x}{\partial \bar{z}} + \frac{\partial v}{\partial y}\frac{\partial y}{\partial \bar{z}}\right)\\ &= \frac{1}{2}\left(\frac{\partial u}{\partial x} + i\frac{\partial u}{\partial y} + i\frac{\partial v}{\partial x} - \frac{\partial v}{\partial y}\right),\end{aligned}$$

which is zero if the Cauchy–Riemann equations are satisfied. Hence, a necessary condition for analyticity is that $w = f(z)$ is independent of $\bar{z}$ and is a function of z alone. It is now clear that $f(z) = z\bar{z}$ cannot be analytic, and in any formula representing an analytic function, x and y can occur only in the combination $x + iy$.

Conjugate functions

If $w = u + iv = f(z)$ where $f(z)$ is analytic in some domain, the real functions u and v of the two real variables x and y are called *conjugate functions*. Note that the word "conjugate" here has nothing to do with the complex conjugate of a number.

Conjugate functions satisfy the Cauchy–Riemann equations (5.4), and by taking the x derivative of the first and the y derivative of the second, we have

$$\frac{\partial^2 u}{\partial x^2} = \frac{\partial^2 v}{\partial y\,\partial x}, \qquad \frac{\partial^2 u}{\partial y^2} = -\frac{\partial^2 v}{\partial x\,\partial y}.$$

It is shown later that if $f(z)$ is analytic, its derivatives of all orders exist and are continuous, ensuring that the second partial derivatives of u and v are continuous. Hence

$$\frac{\partial^2 v}{\partial y\,\partial x} = \frac{\partial^2 v}{\partial x\,\partial y},$$

giving

$$\frac{\partial^2 u}{\partial x^2} + \frac{\partial^2 u}{\partial y^2} = 0, \tag{5.9}$$

and v also satisfies the same equation.

Equation (5.9) occurs frequently in mathematical physics and is *Laplace's equation* in two dimensions. It is, for example, satisfied by a planar electrostatic potential at a point not occupied by a charge, and by the velocity potential and stream function for the two-dimensional irrotational flow of an incompressible inviscid fluid. By separating any analytic function of z into its real and imaginary parts we obtain immediately two solutions of Laplace's equation, and because of this, the theory of functions of a complex variable has important applications to the solution of two-dimensional problems. Conversely, any solution of Laplace's equation having continuous second partial derivatives is a conjugate function.

Given one of two conjugate functions, the Cauchy–Riemann equations can be used to find the other. It is therefore possible to construct an analytic function from a knowledge of its real or imaginary part in some domain.

Example. Show that

$$u(x, y) = y^3 - 3x^2 y$$

is the real part of a function analytic in some domain. Deduce the conjugate function $v(x, y)$ and construct the analytic function $f(z) = u + iv$.

It is easily verified that u satisfies Laplace's equation and has continuous second partial derivatives, thereby completing the first part of the problem. If v is the conjugate function,

$$\frac{\partial u}{\partial x} = -6xy = \frac{\partial v}{\partial y}$$

from the first Cauchy–Riemann equation, and by integrating with respect to y we find

$$v = -3xy^2 + g_1(x), \tag{5.10}$$

where $g_1(x)$ is an arbitrary function of x. Similarly

$$\frac{\partial u}{\partial y} = 3y^2 - 3x^2 = -\frac{\partial v}{\partial x}$$

from the second Cauchy–Riemann equation, and by integrating with respect to x we find

$$v = -3xy^2 + x^3 + g_2(y), \tag{5.11}$$

where $g_2(y)$ is an arbitrary function of y. Comparison of (5.10) and (5.11) shows

$$g_1(x) = x^3 + c, \qquad g_2(y) = c,$$

where c is an arbitrary constant, and hence

$$v(x, y) = x^3 - 3xy^2 + c.$$

The corresponding analytic function is

$$f(z) = y^3 - 3x^2y + i(x^3 - 3xy^2 + c). \tag{5.12}$$

We know that the right-hand side can be expressed as a function of $z = x + iy$, and we could obtain the required form by inserting the expressions (5.8) for x and y and simplifying. The terms involving $\bar{z}$ must cancel out. Alternatively, since every x must be associated with a z and since (5.12) reduces to $i(x^3 + c)$ when $y = 0$, we expect that

$$f(z) = i(z^3 + c),$$

and it can be verified that this is so.

An alternative and very elegant method for reconstructing an analytic function has been proposed by A. Oppenheim (*Math. Gazette* 36 (1952) pp. 286–7) and this avoids the need for differentiation and integration. If $u(x, y)$ is known, the expressions (5.8) for x and y can be inserted to give

$$u\left\{\frac{1}{2}(z + \bar{z}), \frac{1}{2i}(z - \bar{z})\right\} = \phi(z) + \psi(\bar{z}) \quad \text{(say)}.$$

By using the Cauchy–Riemann equations, the conjugate function is found to be

$$v(x, y) = -i\{\phi(z) - \psi(\bar{z}) + c\},$$

where c is an arbitrary constant (see Exercise 5, following), and hence

$$f(z) = u + iv = 2\phi(z) + c.$$

As an example, if

$$u(x, y) = e^x \cos y,$$

we have

$$u(x, y) = \tfrac{1}{2}e^x(e^{iy} + e^{-iy}) = \tfrac{1}{2}(e^z + e^{\bar{z}}),$$

giving

$$f(z) = e^z + c.$$

It must be emphasized that not all real functions of x and y are the real parts of analytic functions, and not all functions of a complex variable are analytic in some domain. Because of the constraints on $u(x, y)$ and $v(x, y)$, analytic functions are only a small subset of the general class of functions of z, but they are the ones of practical interest, and most functions we will encounter are analytic everywhere or almost everywhere. Simple examples of analytic functions are polynomials and exponentials. The sums and products of analytic functions are analytic, and the quotient of two analytic functions is analytic except at the zeros of the denominator. An analytic function of an analytic function is also analytic.

Exercises

1. Find the derivative $f'(z)$ if it exists, and state where $f(z)$ is analytic.
   *(i) $f(z) = (1 - 3z^2)^4$. (ii) $f(z) = |z|\sin z$.
   (iii) $f(z) = \cos x \cosh y + i \sin x \sinh y$.
   (iv) $f(z) = \dfrac{x + iy}{x^2 + y^2}$.
2. Under what conditions is $v(x, y) = c_1x^3 + c_2x^2y + c_3xy^2 + c_4y^3$ the imaginary part of an analytic function?
3. Construct an analytic function $f(z)$ whose real part is $e^{-x}(x \cos y + y \sin y)$ with $f(0) = 1$.
4. Determine if each of the following functions $u(x, y)$ is the real part of a function analytic in some domain. Where possible, find the

conjugate function $v(x, y)$ and, hence, the analytic function $f(z) = u + iv$.

*(i) $e^x \sin y + y$. (ii) $x^4 - 3x^2y^2$.
(iii) $\cosh x \sin y$. (iv) $x^2 - y^2 - y$.

5. If $u(x, y) = \phi(z) + \psi(\bar{z})$ is the real part of an analytic function:
   (i) Show that

$$\frac{\partial u}{\partial x} = \frac{\partial \phi}{\partial z} + \frac{\partial \psi}{\partial \bar{z}}, \qquad \frac{\partial u}{\partial y} = i\left(\frac{\partial \phi}{\partial z} - \frac{\partial \psi}{\partial \bar{z}}\right).$$

   (ii) Use the Cauchy–Riemann equations to show that the conjugate function is

$$v(x, y) = i\{-\phi(z) + \psi(\bar{z})\} + c,$$

   where c is an arbitrary constant.

6. If u is the real part of a function $f(z)$ analytic in some domain, use Oppenheim's method to find $f(z)$ for the following $u(x, y)$.
   *(i) $2x(1 - y)$. (ii) $x - e^{-y}\sin x$.
   (iii) $\dfrac{y}{x^2 + y^2}$. (iv) $\sinh x \sin y$.

7. In fluid mechanics the velocity potential ϕ and the stream function ψ are such that $f(z) = \phi + i\psi$ where $f(z)$ is an analytic function. If $\phi = x^2 + 4x - y^2 + 2y$, find $f(z)$.

8. If $f(z) = u + iv$ is analytic in some domain, show that the families of curves $u =$ constant and $v =$ constant are orthogonal; that is, show that the curves $u = c_1$ and $v = c_2$ intersect at right angles provided $f'(z) \neq 0$ at the point of intersection.

9. If $f(z) = u + iv = z^2$, sketch the families of curves $u =$ constant and $v =$ constant. Are they in accord with the previous exercise, including the curves for $u = 0$ and $v = 0$?

10. If $f(z) = u + iv$ is analytic in some domain, show that

$$\left(\frac{\partial^2}{\partial x^2} + \frac{\partial^2}{\partial y^2}\right)(u^2 + v^2) = 4\left\{\left(\frac{\partial u}{\partial x}\right)^2 + \left(\frac{\partial v}{\partial x}\right)^2\right\}$$

and deduce that

$$\left(\frac{\partial^2}{\partial x^2} + \frac{\partial^2}{\partial y^2}\right)|f(z)|^2 = 4|f'(z)|^2.$$

5.3 Integration in the complex plane

Analyticity was defined in terms of the existence of a unique derivative, but in following Cauchy's original development of complex-variable theory, further progress depends on the representation of a function as an integral in the complex plane. The integrals in question are line integrals. These have no physical meaning as such, and it is necessary to resist the temptation to seek a physical interpretation of them.

Line integrals

A *contour*, *path*, or *line* (we shall use the terms interchangeably) in the complex plane is the set of points $z(t) = x(t) + iy(t)$ where $a \leq t \leq b$ is a real parameter. It will be assumed that $x(t)$ and $y(t)$ are piecewise-smooth functions of t in the sense defined in Section 4.5; that is, $x(t)$ and $y(t)$ are continuous with piecewise-continuous derivatives. A contour can therefore have kinks or corners, but two disjoint lines together do not form a contour. In short, it must be possible to draw it without taking the pencil from the paper.

If $f(z) = u + iv$ is a complex function of z not necessarily analytic, and C is a contour in the complex plane with $z = z(t)$, $a \leq t \leq b$ on C, the integral of $f(z)$ along C is written as

$$\int_C f(z)\,dz,$$

where the integration is in a specified *direction* on C which can be indicated by an arrow on the contour and is usually in the direction of increasing t. The mathematical definition is similar to that of a real line integral. The range $a \leq t \leq b$ is subdivided into n intervals by t_0 $(= a)$, $t_1, t_2, \ldots, t_n$ $(= b)$ as shown in Fig. 5.7. Let $z_j = x(t_j) + iy(t_j)$, $\Delta z_j = z_j - z_{j-1}$, $\Delta t_j = t_j - t_{j-1}$, and choose an arbitrary value t_j^* in the interval $t_{j-1} \leq t \leq t_j$ with $z_j^* = x(t_j^*) + iy(t_j^*)$. Then

$$\int_C f(z)\,dz = \lim_{\substack{n\to\infty \\ \max \Delta t_j \to 0}} \sum_{j=1}^{n} f(z_j^*)\,\Delta z_j. \tag{5.13}$$

By separating the right-hand side into real and imaginary parts, we obtain

$$\begin{aligned}\int_C f(z)\,dz &= \lim \Sigma(u + iv)(\Delta x + i\,\Delta y) \\ &= \lim\{\Sigma(u\,\Delta x - v\,\Delta y) + i\Sigma(v\,\Delta x + u\,\Delta y)\},\end{aligned}$$

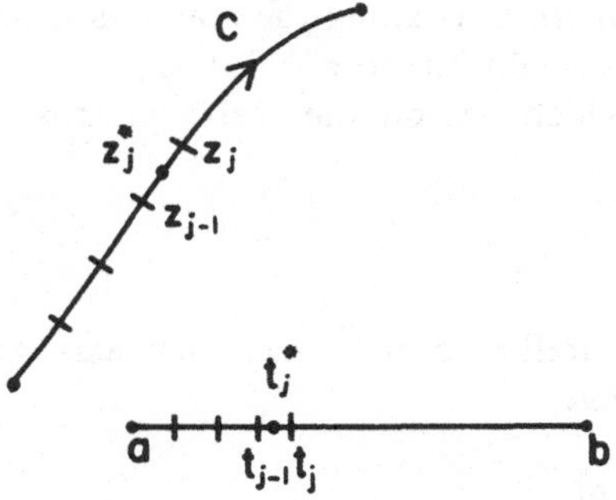

Fig. 5.7: Subdivision of the contour C.

giving

$$\int_C f(z)\,dz = \int_C (u\,dx - v\,dy) + i\int_C (v\,dx + u\,dy). \tag{5.14}$$

Thus, a complex line integral is simply a combination of two real line integrals, and the usual results for real line integrals are immediately applicable. In particular, if C consists of a finite number of segments C_r,

$$\int_C f(z)\,dz = \sum_r \int_{C_r} f(z)\,dz,$$

and if the direction of integration along C is reversed, the integral is multiplied by -1. Using the parameterization inherent in the definition of C, each real line integral in (5.14) can be converted into a definite integral and hence evaluated. We have

$$\int_C f(z)\,dz = \int_a^b \left(u\frac{dx}{dt} - v\frac{dy}{dt}\right) dt + i\int_a^b \left(v\frac{dx}{dt} + u\frac{dy}{dt}\right) dt,$$

which can be written more concisely as

$$\int_C f(z)\,dz = \int_a^b f\{z(t)\}\frac{dz}{dt}\,dt. \tag{5.15}$$

Example 1. Evaluate $\int_C z\,dz$ where C is the arc of the parabola $y^2 = 4x$ from the origin to $z = 1 + 2i$.

A convenient parameterization of C is $x = t^2$, $y = 2t$ with $0 \le t \le 1$. Hence $z = t^2 + 2it$ and since $dz/dt = 2(t + i)$, we have

$$\begin{aligned}\int_C z\,dz &= \int_0^1 (t^2 + 2it)2(t + i)\,dt \\ &= \int_0^1 (2t^3 + 6it^2 - 4t)\,dt = -\tfrac{3}{2} + 2i.\end{aligned}$$

Example 2. Evaluate $\int_C z^2\,dz$ where C is along the real axis from $z = 0$ to $z = 1$ and thence along the straight line to $z = 1 + i$.

For the portion C_1 of C which lies on the real axis, $z = t$ where $0 \le t \le 1$ and

$$\int_C z^2\,dz = \int_0^1 t^2\,dt = \tfrac{1}{3}.$$

For the portion C_2 that is parallel to the imaginary axis we write $z = 1 + it$ with $0 \le t \le 1$ so that

$$\int_C z^2\,dz = \int_0^1 (1 + it)^2 i\,dt$$

$$= i\int_0^1 (1 + 2it - t^2)\,dt = -1 + \tfrac{2}{3}i,$$

and hence

$$\int_C z^2\,dz = \int_{C_1} z^2\,dz + \int_{C_2} z^2\,dz = -\tfrac{2}{3}(1 - i).$$

It can be verified that in each of these examples the same result is obtained by integration along a straight-line path between the endpoints. This illustrates the fact that in some instances a line integral is independent of the path, but until the conditions for this are established, the integration should be carried out along the path that is given.

Example 3. Evaluate $\int_C z^n\,dz$ where n is an integer, positive, negative, or zero, and C is a circle of unit radius, centered on the origin, and described anticlockwise.

The standard parameterization of a circle of unit radius is

$$x = \cos t, \qquad y = \sin t,$$

where t is the angle subtended at the center. As the circle is described anticlockwise starting at the point $x = 1$, $y = 0$, the angle t ranges from 0 to 2π. The corresponding parameterization of z is then

$$z = \cos t + i \sin t = e^{it}, \qquad 0 \le t \le 2\pi.$$

Since $dz/dt = ie^{it}$, we have

$$\int_C z^n\,dz = \int_0^{2\pi} e^{int} ie^{it}\,dt = i\int_0^{2\pi} e^{i(n+1)t}\,dt.$$

If $n \ne -1$,

$$\int_0^{2\pi} e^{i(n+1)t}\,dt = \left[\frac{e^{i(n+1)t}}{i(n+1)}\right]_0^{2\pi} = 0$$

and hence,

$$\int_C z^n \, dz = 0, \qquad n \neq -1, \tag{5.16}$$

whereas if $n = -1$,

$$\int_C \frac{1}{z} \, dz = i \int_0^{2\pi} dt = 2\pi i. \tag{5.17}$$

In the last example the path of integration is closed, and when this is so we shall add a circle to the integral sign to emphasize the fact. The direction in which C is described will be shown by an arrow on the circle.

It is sometimes necessary to obtain an upper bound for the magnitude of a line integral, and we can do this as follows. From the definition (5.13) and the triangle inequality (1.8),

$$\begin{aligned}\left|\int_C f(z)\, dz\right| &= \lim_{\substack{n\to\infty \\ \max \Delta t_j \to 0}} \left|\sum_{j=1}^{n} f(z_j^*)\,\Delta z_j\right| \\ &\leq \lim_{\substack{n\to\infty \\ \max \Delta t_j \to 0}} \sum_{j=1}^{n} \left|f(z_j^*)\,\Delta z_j\right| \\ &= \lim_{\substack{n\to\infty \\ \max \Delta t_j \to 0}} \sum_{j=1}^{n} \left|f(z_j^*)\right| \left|\Delta z_j\right|,\end{aligned}$$

giving

$$\left|\int_C f(z)\, dz\right| \leq \int_C |f(z)|\, |dz|. \tag{5.18}$$

But

$$\int_C |dz| = \int_a^b \left\{\left(\frac{dx}{dt}\right)^2 + \left(\frac{dy}{dt}\right)^2\right\}^{1/2} dt = \int_C ds = L,$$

where ds is an element of arclength and L is the length of C. Hence, if $|f(z)|$ is bounded on C (i.e. $|f(z)| < M$),

$$\left|\int_C f(z)\, dz\right| \leq M \int_C dz = ML. \tag{5.19}$$

Green's theorem

This theorem in real variable theory relates a double integral in the xy plane to a line integral around the boundary of the region. It is one of

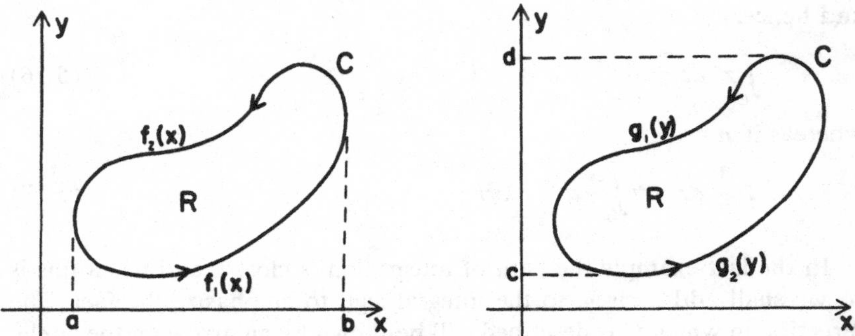

Fig. 5.8: An example of the special region used for Green's theorem.

several similar results discovered by the English mathematician George Green (1793–1841) and included in a memoir privately circulated in 1828.

Theorem: If R is a closed bounded region in the xy plane whose boundary C consists of a finite number of piecewise-smooth curves, and $P(x, y)$ and $Q(x, y)$ are continuous with continuous partial derivatives $\partial P/\partial y$ and $\partial Q/\partial x$ everywhere in a domain D containing R, then

$$\iint_R \left(\frac{\partial Q}{\partial x} - \frac{\partial P}{\partial y} \right) dx\, dy = \oint_C (P\, dx + Q\, dy), \tag{5.20}$$

where the integration along C is such that R remains on the left as one advances in the direction of integration.

Proof: The theorem is proved first for the case in which R is a special region that can be represented in both of the forms

$$a \le x \le b, \qquad f_1(x) \le y \le f_2(x),$$

$$c \le y \le d, \qquad g_1(y) \le x \le g_2(y)$$

as shown in Fig. 5.8.

The double integral $\iint_R (\partial P/\partial y)\, dx\, dy$ can be written as a repeated integral:

$$\iint_R \frac{\partial P}{\partial y}\, dx\, dy = \int_a^b \int_{f_1(x)}^{f_2(x)} \frac{\partial P}{\partial y}\, dy\, dx;$$

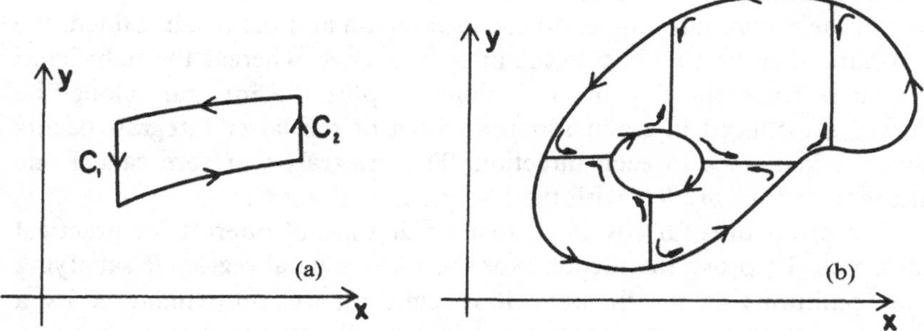

Fig. 5.9: Regions used in the proof of Green's theorem.

the inner integral can be evaluated to give

$$\iint_R \frac{\partial P}{\partial y}\,dx\,dy = \int_a^b \{P[x, f_2(x)] - P[x, f_1(x)]\}\,dx$$

$$= -\int_b^a P[x, f_2(x)]\,dx - \int_a^b P[x, f_1(x)]\,dx$$

$$= -\oint_C P\,dx.$$

If any portions of C are parallel to the y axis, such as the segments C_1 and C_2 in Fig. 5.9(a), the integrals over these portions are zero and do not affect the preceding result. Similarly

$$\iint_R \frac{\partial Q}{\partial x}\,dx\,dy = \int_c^d \int_{g_1(y)}^{g_2(y)} \frac{\partial Q}{\partial x}\,dx\,dy$$

$$= \int_c^d Q[g_2(y), y]\,dy + \int_d^c Q[g_1(y), y]\,dy$$

$$= \oint_C Q\,dy,$$

and hence

$$\iint_R \left(\frac{\partial Q}{\partial x} - \frac{\partial P}{\partial y}\right) dx\,dy = \oint_C (P\,dx + Q\,dy),$$

which completes the proof for the special type of region.

Suppose next that R itself is not of this type, but can be divided into a finite number of such regions by the addition of suitable lines or arcs as in Fig. 5.9(b). Note that the inner boundary of R is described in a clockwise direction in accordance with the statement of the theorem. When the theorem is applied to each subregion and the results added, the left-hand members add up to the integral over R, whereas the right-hand members form the line integral along C plus the integrals along the curves introduced in subdividing R. Each of the latter integrals occurs twice, taken once in each direction. The integrals therefore cancel one another and we are left with the line integral along C.

The proof thus far covers all regions that are of interest for practical purposes. To prove the theorem for the most general region R satisfying the conditions of the theorem, it is sufficient to approximate R by a region of the type considered and then use a limiting process. □

Example. In Eq. (5.20) let $P = 0$ and $Q = x$. Then

$$\iint_R dx\,dy = \oint_C x\,dy$$

and the integral on the left is the area A of R. Similarly, if $P = -y$ and $Q = 0$,

$$\iint_R dx\,dy = -\oint_C y\,dx,$$

and by adding the two formulas we obtain

$$A = \frac{1}{2}\oint_C (x\,dy - y\,dx).$$

This expression for the area of a plane region in terms of a line integral around the boundary is the basis for the design of certain planimeters.

Exercises

1. If C is the square having vertices $(0,0)$, $(1,0)$, $(1,1)$, and $(0,1)$ in the complex plane, evaluate $\oint_C f(z)\,dz$ where $f(z) = u + iv$ and:
   *(i) $u = x,\ v = -y$. (ii) $u = x,\ v = y$.
2. Evaluate $\int_C ze^{-z}\,dz$ where C is the straight line segment from $z = 1$ to $z = 2i$.
3. Evaluate $\int_C \cos z\,dz$ where C is the straight line segment from $z = \pi i$ to $z = 2\pi i$.
4. Evaluate $\int_C \sin z\,dz$ where C is the straight line segment from $z = 0$ to $z = 2i$.
5. Evaluate $\oint_C f(z)\,dz$ where $f(z) = (z^2 + 2z + 3)/z$ and C is the circle $|z| = 3$.

6. Evaluate $\int_C f(z)\,dz$ where C is that part of the circle $|z| = 1$ lying in the upper half-plane, described anticlockwise, and:
*(i) $f(z) = z$. (ii) $f(z) = |z|$.
(iii) $f(z) = 1/z$.

7. Evaluate $\int_C (1/z^2)\,dz$ where C is the arc of the parabola $2y^2 = x + 1$ from $z = 1 + i$ to $z = 1 - i$.

8. Evaluate $\int_C z\,dz$ where C is that part of the ellipse $x^2 + 4y^2 = 1$ lying in the first quadrant, described anticlockwise.

9. If $I = \oint_C e^z/(z+1)^2\,dz$ where C is the circle $|z| = 3$, use Eq. (5.19) to show that $|I| \leq 3\pi e^3/2$.

10. If $I = \int_C z\,dz$ where C is the line segment from 0 to $1 + 2i$, find an upper bound for $|I|$.

5.4 Cauchy's integral theorem and formulas

This theorem is fundamental to the theory of analytic functions and is the foundation for all that follows.

Cauchy's integral theorem

If $f(z)$ is analytic inside and on a closed contour C, then

$$\oint_C f(z)\,dz = 0. \tag{5.21}$$

Proof: From Eq. (5.14) with $f(z) = u + iv$,

$$\oint_C f(z)\,dz = \oint_C (u\,dx - v\,dy) + i\oint_C (v\,dx + u\,dy).$$

If $f'(z)$ is continuous inside and on C, $\partial u/\partial y$ and $\partial v/\partial x$ are also continuous (see Eq. (5.5)) and the conditions of Green's theorem are satisfied with $P = u$ and $Q = -v$. Hence

$$\oint_C (u\,dx - v\,dy) = -\int\!\!\int_R \left(\frac{\partial v}{\partial x} + \frac{\partial u}{\partial y}\right) dx\,dy = 0$$

from the second Cauchy–Riemann equation. Similarly, by using Green's theorem with $P = v$ and $Q = u$,

$$\oint_C (v\,dx + u\,dy) = \int\!\!\int_R \left(\frac{\partial u}{\partial x} - \frac{\partial v}{\partial y}\right) dx\,dy = 0$$

from the first Cauchy–Riemann equation, and therefore

$$\oint_C f(z)\,dz = 0. \quad \square$$

The preceding proof is similar to that originally given by Cauchy, and requires the assumption that $f'(z)$ is continuous in order to justify the application of Green's theorem. As shown later by the French mathematician Edouard Goursat (1858–1936), the assumption is unnecessary, and Cauchy's theorem holds if $f'(z)$ merely exists at all points inside and on C, that is if $f(z)$ is analytic. In fact, the continuity of $f'(z)$ as well as its repeated differentiability are *consequences* of the theorem.

There is an alternative statement of Cauchy's (or, more strictly, the Cauchy–Goursat) integral theorem which refers to a simple (i.e., nonintersecting) closed curve in a simply connected domain. A domain D is *simply connected* if every simple, closed curve in D encloses only points of D. Crudely expressed, the domain has no holes in it, and a domain that is not simply connected (and has holes) is said to be *multiply connected*. The other statement of the theorem is:

If $f(z)$ is analytic in a simply connected domain D, then

$$\oint_C f(z)\, dz = 0 \tag{5.22}$$

for *every* simple closed curve C in D.

The proof is identical to that given previously, and the necessity for having the domain simply connected is evident from the following example.

Example. Evaluate $\oint_C f(z)\, dz$ with $f(z) = 1/z$ where C is the circle $|z| = 1$.

Because of the singularity of $f(z)$ at the point $z = 0$ inside C, the domain of analyticity is the annulus $\varepsilon \leq |z| \leq \infty$ for any $\varepsilon > 0$, and since this is not simply connected, the theorem is not applicable. Indeed, from Eq. (5.17),

$$\oint_C \frac{1}{z}\, dz = 2\pi i.$$

On the other hand, if the contour C were the circle $|z - \frac{3}{2}| = 1$, the origin would lie outside C. Since $f(z)$ is now analytic inside and on C, the integral is zero from the original statement of the theorem – a result that is also consistent with the second statement of the theorem if the domain of analyticity that is considered is simply connected and does not include the origin $z = 0$. In other words, the theorem does not demand that we choose D to be the largest possible domain.

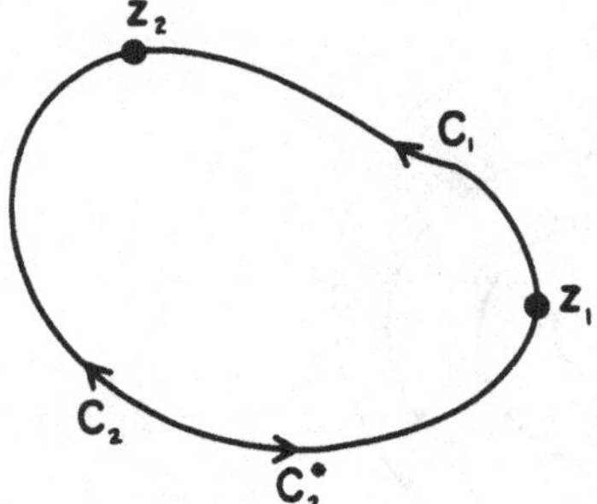

Fig. 5.10: Illustration of path independence.

For many applications the first statement of the theorem is more helpful, but some of the consequences are more easily deduced from the second. In particular, if we subdivide the path C into two arcs C_1 and C_2^* as indicated in Fig. 5.10, Eq. (5.21) becomes

$$\oint_C f(z)\,dz = \int_{C_1} f(z)\,dz + \int_{C_2^*} f(z)\,dz = 0, \tag{5.23}$$

and if C_2 is C_2^* reversed in direction, so that

$$\int_{C_2} f(z)\,dz = -\int_{C_2^*} f(z)\,dz,$$

Eq. (5.23) implies

$$\int_{C_1} f(z)\,dz = \int_{C_2} f(z)\,dz$$

for *any* paths C_1 and C_2 joining the points z_1 and z_2 in a simply connected domain D where $f(z)$ is analytic. The integral is therefore *path-independent* provided the paths are confined to the domain D, and its value is a function only of $f(z)$ and the endpoints, that is,

$$\int_{C_1} f(z)\,dz = \int_{z_1}^{z_2} f(z)\,dz.$$

As an example, if $f(z) = g'(z)$, then

$$\int_{z_1}^{z_2} f(z)\,dz = [g(z)]_{z_1}^{z_2} = g(z_2) - g(z_1),$$

which is the same as in elementary calculus. Thus

$$\int_1^i z^2\,dz = \left[\frac{z^3}{3}\right]_1^i = -\frac{1}{3}(1+i)$$

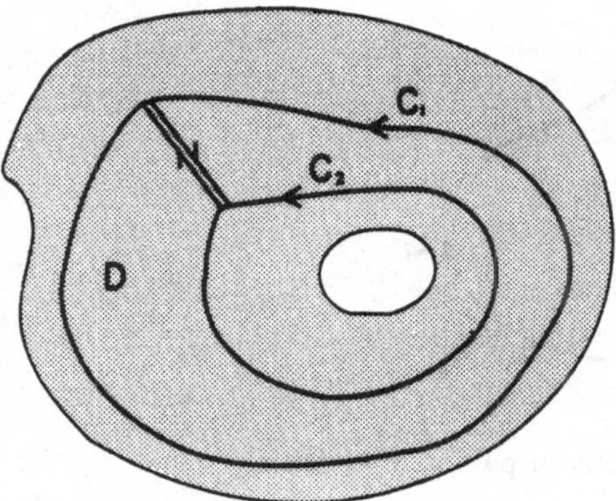

Fig. 5.11: Illustration of path deformation.

for any path, and

$$\int_{-i}^{i} \frac{1}{z^2}\, dz = \left[-\frac{1}{z} \right]_{-i}^{i} = 2i$$

for any path not through the origin.

Because of the path independence in a simply connected domain of analyticity, we are seldom concerned with the precise details of the path of integration. For many of the exercises at the end of the last section, the answer could have been obtained by using a simpler path between the endpoints. If the original and the simpler path together form a closed contour C with $f(z)$ analytic inside and on C, the use of the simpler path is justified by Cauchy's theorem. The same *principle of path deformation* is also applicable to closed paths in a simply or multiply connected domain, provided the integral is analytic in the region between the paths. If $f(z)$ is analytic in the domain D shown in Fig. 5.11,

$$\oint_{C_1} f(z)\, dz = \oint_{C_2} f(z)\, dz. \tag{5.24}$$

The result follows immediately from the first statement of the theorem by making a crosscut joining any point of C_1 to a point of C_2 to form a closed contour Γ within which $f(z)$ is analytic. Then

$$\oint_{\Gamma} f(z)\, dz = 0,$$

and since the crosscut is traversed twice, once in each direction, its

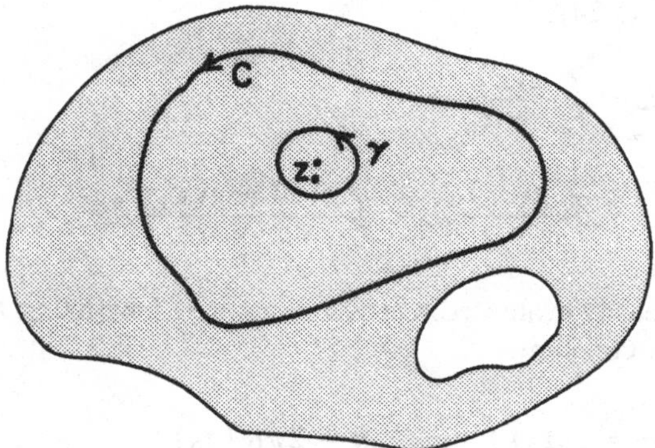

Fig. 5.12: Cauchy's integral formula.

contribution is zero. Hence

$$\oint_{C_1} f(z)\, dz = -\oint_{C_2} f(z)\, dz = \oint_{C_2} f(z)\, dz.$$

When dealing with analytic functions, it is often helpful to visualize a contour as a rubber band, with the boundaries of the domain of analyticity constituting a barrier that cannot be crossed. For an integral between two points z_1 and z_2, the piece of rubber can be pushed hither and yon without affecting the value of the integral; and similarly for a closed contour. Of course, the rubber must not be cut.

Cauchy's integral formula

If $g(z)$ is analytic inside and on a closed contour C, and if z_0 is a point inside C, then

$$g(z_0) = \frac{1}{2\pi i} \oint_C \frac{g(z)}{z - z_0}\, dz. \tag{5.25}$$

Proof: The domain of analyticity of $g(z)$ is not required to be simply connected, but since $g(z)$ is analytic inside and on C, the formula concerns only the simply connected part shown in Fig. 5.12. To prove it, let γ be a small circle of radius r about $z = z_0$ lying wholly in D. In the region between C and γ the function $f(z) = [g(z)]/(z - z_0)$ is analytic and hence, from Eq. (5.24),

$$\oint_C \frac{g(z)}{z-z_0}\,dz = \oint_\gamma \frac{g(z)}{z-z_0}\,dz$$

$$= \oint_\gamma \frac{g(z_0)}{z-z_0}\,dz + \oint_\gamma \frac{g(z)-g(z_0)}{z-z_0}\,dz.$$

The first integral is easily evaluated. On γ, $z - z_0 = re^{it}$ with $0 \le t \le 2\pi$, and since $g(z_0)$ is a constant,

$$\oint_\gamma \frac{g(z_0)}{z-z_0}\,dz = g(z_0)\int_0^{2\pi} \frac{rie^{it}}{re^{it}}\,dt = 2\pi i g(z_0).$$

To evaluate the second integral we note that because $g(z)$ is continuous everywhere inside C, it follows that for any $\varepsilon > 0$ there exists a δ such that $|g(z) - g(z_0)| < \varepsilon$ if $|z - z_0| = r < \delta$. Equation (5.19) then gives

$$\left|\oint_\gamma \frac{g(z)-g(z_0)}{z-z_0}\,dz\right| < \frac{\varepsilon}{r}2\pi r = 2\pi\varepsilon,$$

which can be made as small as we like by choosing r sufficiently small. Since the integral has the same value for all r, the only possibility is that the integral is zero, completing the proof. □

The integral formula (5.25) shows that an analytic function is completely determined throughout the interior of C by its values on C, but even more remarkable is the fact that all derivatives are likewise determined by the boundary values of the function. This striking result has no analog in real variable theory, and is proved next.

Cauchy's general integral formula

If $g(z)$ is analytic inside and on a closed contour C, and if z_0 is a point inside C, then

$$g^{(n)}(z_0) = \frac{n!}{2\pi i}\oint_C \frac{g(z)}{(z-z_0)^{n+1}}\,dz \qquad (5.26)$$

for $n = 0, 1, 2, \ldots$.

Thus, $g(z)$ has derivatives of all orders at *every* point z_0 inside C, with values that are given by Eq. (5.26).

Proof: The case $n = 0$ is the integral formula (5.25) and requires no further discussion, and we shall consider first $n = 1$. From Eq. (5.25)

$$\begin{aligned}\frac{g(z_0 + h) - g(z_0)}{h} &= \frac{1}{2\pi ih}\oint_C g(z)\left\{\frac{1}{z - z_0 - h} - \frac{1}{z - z_0}\right\} dz \\ &= \frac{1}{2\pi i}\oint_C \frac{g(z)}{(z - z_0)(z - z_0 - h)}\, dz \\ &= \frac{1}{2\pi i}\oint_C \frac{g(z)}{(z - z_0)^2}\, dz + I,\end{aligned}$$

where

$$\begin{aligned}I &= \frac{1}{2\pi i}\oint_C \frac{g(z)}{z - z_0}\left\{\frac{1}{z - z_0 - h} - \frac{1}{z - z_0}\right\} dz \\ &= \frac{h}{2\pi i}\oint_C \frac{g(z)}{(z - z_0)^2(z - z_0 - h)}\, dz,\end{aligned}$$

and if we can prove that $|I| \to 0$ as $h \to 0$, the required result is established. Since $g(z)$ is analytic inside and on C, it is certainly bounded, so that $|g(z)| \le M$ on C. Let d be the lower bound of the distance of z_0 from C. Then $|z - z_0| \ge d$ for all z on C, and if h is so small that $|h| < \frac{1}{2}d$, we have $|z - z_0 - h| > \frac{1}{2}d$ from the triangle inequality (1.9). Equation (5.19) now gives

$$|I| < \frac{|h|}{2\pi}\frac{ML}{d^2 \cdot d/2},$$

where L is the length of C, and this tends to zero as $|h| \to 0$.

If we assume the formula is proved for $n = m$ and consider the expression

$$\frac{g^{(m)}(z_0 + h) - g^{(m)}(z_0)}{h},$$

it is readily shown that this equals

$$\frac{(m + 1)!}{2\pi i}\oint_C \frac{g(z)}{(z - z_0)^{m+2}}\, dz + I,$$

and the proof that $|I|$ tends to zero as $|h| \to 0$ can be carried out as before. Since the formula is valid for $n = 0$ and $n = 1$, the general result now follows by induction. □

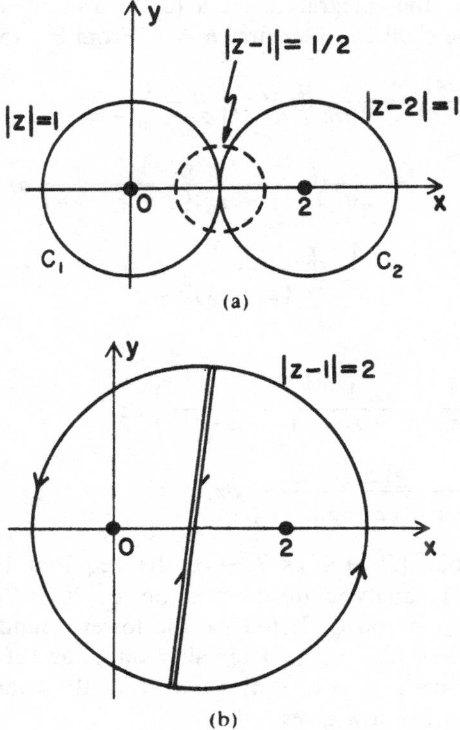

Fig. 5.13: Evaluation of the integral I in the example.

Cauchy's general integral formula is a powerful tool for evaluating integrals around closed contours, and as an aid to memory, we observe that Eq. (5.26) is equivalent to differentiating (5.25) n times with respect to z_0 under the integral sign. Integration has now been reduced to a process of differentiation at worst.

Example. Evaluate $I = \oint_C f(z)\,dz$ where $f(z) = 4/[z^2(z-2)]$ and the contour C is:

(i) $|z-1| = \frac{1}{2}$. (ii) $|z-2| = 1$.
(iii) $|z| = 1$. (iv) $|z-1| = 2$.

(i) The function $f(z)$ is singular ("blows up") at the points $z = 0$ and $z = 2$ shown in Fig. 5.13 (a), and the contour $|z-1| = \frac{1}{2}$ is a circle of radius $\frac{1}{2}$ centered on $z = 1$. Both singularities are outside the contour C, and since $f(z)$ is therefore analytic inside and on C,

$$I = 0$$

by Cauchy's integral theorem.

(ii) The contour $|z - 2| = 1$ is a circle of unit radius centered on $z = 2$, and the singularity at $z = 2$ is included, but the one at $z = 0$ is not. Hence, if

$$f(z) = \frac{g(z)}{z-2} \quad \text{with} \quad g(z) = \frac{4}{z^2},$$

$g(z)$ is analytic inside and on C, and from Cauchy's integral formula (5.25) with $z_0 = 2$, we have

$$I = 2\pi i g(2) = 2\pi i.$$

(iii) The contour $|z| = 1$ encloses the singularity at $z = 0$ but not the one at $z = 2$. If

$$f(z) = \frac{g(z)}{z^2} \quad \text{with} \quad g(z) = \frac{4}{z-2},$$

$g(z)$ is analytic inside and on C, and from Cauchy's general integral formula (5.26) with $n = 1$ and $z_0 = 0$,

$$I = \frac{2\pi i}{1!} g'(0) = -2\pi i.$$

(iv) The contour $|z - 1| = 2$ encloses both singularities, and there are now two ways that we can proceed. By expanding $f(z)$ in partial fractions we obtain

$$f(z) = \frac{1}{z-2} - \frac{2}{z^2} - \frac{1}{z},$$

leading to the decomposition

$$I = I_1 + I_2 + I_3,$$

where

$$I_j = \oint_C f_j(z)\, dz$$

with

$$f_1(z) = \frac{1}{z-2}, \qquad f_2(z) = -\frac{2}{z^2}, \qquad f_3(z) = -\frac{1}{z}.$$

Since each integrand has only one singularity inside C (or, indeed, anywhere), the formulas are applicable to the individual integrals. For I_1 we have $g(z) = 1$ and $z_0 = 2$, so that $I_1 = 2\pi i$. For I_2, $g(z) = -2$,

$n = 1$, and $z_0 = 0$, and because $g'(z) = 0$, $I = 0$. Finally, for I_3, $g(z) = -1$ and $z_0 = 0$, giving $I_3 = -2\pi i$, and when these are substituted into the expression for I we have

$$I = 2\pi i + 0 - 2\pi i = 0,$$

in spite of the fact that there are two singularities within the contour.

An alternative and more direct procedure makes use of the principle of path deformation. Since $f(z)$ is analytic in the region between the contour C and, for example, the contour formed by the (touching) circles $|z| = 1$ and $|z - 2| = 1$, it follows immediately that

$$I = \oint_{C_1} g(z)\,dz + \oint_{C_2} f(z)\,dz,$$

where C_1 and C_2 are the circles $|z| = 1$ and $|z - 2| = 1$, respectively. This eliminates the need for a partial-fraction expansion. Only one singularity is associated with each integral, and from the solutions of (ii) and (iii),

$$I = 2\pi i - 2\pi i = 0.$$

As evident from Fig. 5.11, the process is equivalent to the introduction of a cut joining two points of C to form two contours, each enclosing one singularity. This is shown in Fig. 5.13(b), and since the cut is traversed twice, once in each direction, it has no effect on the value of the integral.

When using the formulas to evaluate $\oint_C f(z)\,dz$, the steps that are involved are as follows:

(i) Determination of the contour C and the direction in which it is described. A sketch is often helpful, and it should be noted that the contour must be closed.

(ii) Identification of the singularities of $f(z)$ inside C. Singularities that are outside C are of no concern, and if all of them are outside, the integral is zero by Cauchy's integral theorem. If it appears that there is a singularity on C, it is advisable to check again. Since no meaning has been given to integration *through* a singularity, somebody has made a grievous error.

(iii) Determination of $g(z)$, z_0, and n for each singularity inside C, and the computation of the resulting contribution to the integral. From the principle of path deformation or by the introduction of appropriate cuts, the value of the integral is the sum of the contributions from the individual singularities.

Exercises

1. Evaluate $\oint_C f(z)\,dz$ where C is the unit circle $|z| = 1$ and:
*(i) $f(z) = 1/z$. (ii) $f(z) = e^{iz}$.
(iii) $f(z) = 1/(z^2 + 2)$. (iv) $f(z) = 1/(2z + i)$.

2. Repeat the previous exercise with:
(i) $f(z) = (\cos z)/z$. (ii) $f(z) = (\sin z)/z$.
(iii) $f(z) = e^{iz}/z^4$. (iv) $f(z) = (e^z - 1)/z^2$.

3. If $f(z) = (z^2 + 2)^2/[z(z-1)^3]$, evaluate $\oint_C f(z)\,dz$ where C is:
(i) $|z - 2 - i| = \sqrt{3}$. (ii) $|z + 1| = \frac{1}{2}$.
(iii) $|z| = \frac{1}{2}$. (iv) $|z + i| = \sqrt{3}$.

4. If C is the equilateral triangle in the upper-half z plane with base from $z = -R$ to $z = R$ where R is real with $R > 1$ and C is described by first going from $z = -R$ to $z = R$, evaluate $\oint_C f(z)\,dz$ where:
*(i) $f(z) = \dfrac{1}{z^2 + 1}$. (ii) $f(z) = \dfrac{e^{2iz}}{z^2 + 1}$.

5. Repeat the previous exercise with the triangle in the lower half-plane.

6. If C is the contour from $z = -R$ to $z = R$ (R real) along the real axis and then back to $z = -R$ following a semicircle of radius R in the upper half-plane, show

$$\oint_C \frac{dz}{z^2 + 1} = \begin{cases} 0 & \text{if } R < 1, \\ \pi & \text{if } R > 1. \end{cases}$$

7. If C is the semicircular contour defined above with $R > a$ where a is real ($a > 0$), prove that

$$\oint_C \frac{dz}{(z^2 + a^2)^{n+1}} = \frac{2\pi}{(2a)^{2n+1}} \frac{(2n)!}{(n!)^2}$$

for $n = 0, 1, 2, \ldots$.

8. Using Cauchy's general integral formula, show that

$$\left(\frac{a^n}{n!}\right)^2 = \frac{1}{2\pi i} \oint_C \frac{a^n e^{az}}{n! z^{n+1}}\,dz,$$

where C is any closed contour surrounding the origin. Since

$$\sum_{n=0}^{\infty} \frac{t^n}{n!} = e^t,$$

show that

$$\sum_{n=0}^{\infty}\left(\frac{a^n}{n!}\right)^2 = \frac{1}{2\pi}\int_0^{2\pi} e^{2a\cos\theta}\,d\theta.$$

5.5 Power series

Until now the entire theory of analytic functions has been based on the requirement for a unique derivative. Using the general integral formula we have shown that the existence of $f'(z)$ implies the existence of $f''(z)$ and, hence, the continuity of $f'(z)$ throughout the domain, and so on for all derivatives of $f(z)$. The formula also provides a simple means for evaluating certain integrals around closed contours, and the formula is, indeed, sufficient for the immediate task of treating some of the integrals associated with the Fourier and inverse Laplace transforms considered in later chapters. But were we to leave the theory at this stage we would be ignoring the developments made possible by the representation of analytic functions in terms of power series. Although this is a little like changing horses in midstream, the representation leads to a simplification of the results already obtained, and at least a brief discussion is appropriate.

Taylor series

The series of concern to us are power series whose terms are integer powers of $z - z_0$, for example

$$\sum_{n=0}^{\infty} a_n(z - z_0)^n, \tag{5.27}$$

where the coefficients a_n are independent of z, and z_0 is a complex constant. The series is a particular version of those discussed in Section 4.5 and the same comments about convergence apply. By definition the series converges if the sequence of partial sums converges, and in any given case we can determine the region of convergence by computing the partial sums for a variety of z or, more rigorously, by applying one of the standard convergence tests. Consider, for example, the series

$$\sum_{n=0}^{\infty} (z - z_0)^n.$$

From the ratio test it is found that the series converges if $|z - z_0| < 1$ and diverges if $|z - z_0| > 1$. The region of convergence is therefore the interior of a unit circle centered on $z = z_0$. The boundary is the circle $|z - z_0| = 1$ and is termed the *circle of convergence*.

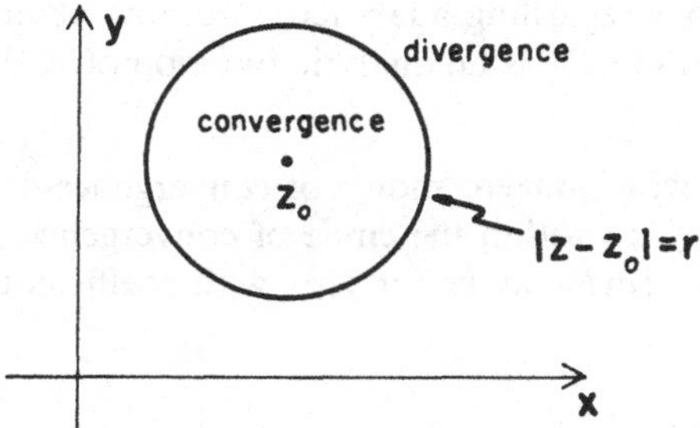

Fig. 5.14: Region of convergence of a power series.

This behavior is typical of the more general power series (5.27). The series converges absolutely and uniformly in some region $|z - z_0| < r$ and diverges for $|z - z_0| > r$ (see Fig. 5.14). The radius r of the circle of convergence can be zero, in which case the series converges only for $z = z_0$; or a positive number, as in the previous example; or infinity, in which case the series converges for all z. No statement can be made about convergence on the circle $|z - z_0| = r$, and the series may converge at some points on the circle or none. Where the series diverges it is meaningless and provides no information, but within the circle of convergence the series defines a function of z, which we shall denote by $f(z)$. We now have

$$f(z) = \sum_{n=0}^{\infty} a_n (z - z_0)^n, \qquad |z - z_0| < r,$$

which resembles the Taylor-series representation or expansion of the function $f(z)$ about the point $z = z_0$. Assuming $r \neq 0$, there is a region where the series converges absolutely and uniformly, allowing term-by-term differentiation. Thus

$$f'(z) = \sum_{n=1}^{\infty} n a_n (z - z_0)^{n-1}$$

so that

$$a_1 = f'(z_0);$$

similarly

$$f''(z) = \sum_{n=2}^{\infty} n(n-1) a_n (z - z_0)^{n-2}$$

giving

$$a_2 = \tfrac{1}{2} f''(z_0),$$

and so on. The implications regarding analyticity are now obvious, and since each term in the series (5.27) is an analytic function of z, the result is hardly surprising:

> any power series with nonzero radius of convergence defines an analytic function $f(z)$ within the circle of convergence, and the power series is the *Taylor series* for $f(z)$ with coefficients
>
> $$a_n = \frac{f^{(n)}(z_0)}{n!}.$$

The series is named after the English mathematician Brook Taylor (1685–1731), who published the real variable counterpart in 1715.

The converse is also true, and any analytic function can be expanded in a Taylor series about a point z_0 in the domain of analyticity. Suppose $f(z)$ is analytic inside and on a circle C of radius r with center at $z = z_0$. For z inside C, Cauchy's integral formula (5.25) gives

$$\begin{aligned} f(z) &= \frac{1}{2\pi i}\oint_C \frac{f(\xi)}{\xi - z}\,d\xi \\ &= \frac{1}{2\pi i}\oint_C \frac{f(\xi)}{\xi - z_0}\left(1 - \frac{z - z_0}{\xi - z_0}\right)^{-1} d\xi. \end{aligned} \tag{5.28}$$

From the geometric progression

$$1 + q + q^2 + \cdots + q^n = \frac{1 - q^{n-1}}{1 - q}, \qquad q \neq 1$$

we obtain

$$(1 - q)^{-1} = 1 + q + q^2 + \cdots, \tag{5.29}$$

and as noted above, the series is uniformly convergent if $|q| < 1$. Let $q = (z - z_0)/(\xi - z_0)$. Since its magnitude is less than unity, (5.29) can be used to expand the corresponding factor in (5.28), and because of the uniform convergence of the series, the integration can be carried out term by term. Hence

$$\begin{aligned} f(z) &= \frac{1}{2\pi i}\oint_C \frac{f(\xi)}{\xi - z_0}\,d\xi + (z - z_0)\frac{1}{2\pi i}\oint_C \frac{f(\xi)}{(\xi - z_0)}\,d\xi + \cdots \\ &\quad + (z - z_0)^n \frac{1}{2\pi i}\oint_C \frac{f(\xi)}{(\xi - z_0)^{n+1}}\,d\xi + \cdots \\ &= \sum_{n=0}^{\infty} (z - z_0)^n \frac{f^{(n)}(z_0)}{n!} \end{aligned}$$

from Cauchy's general integral formula (5.26).

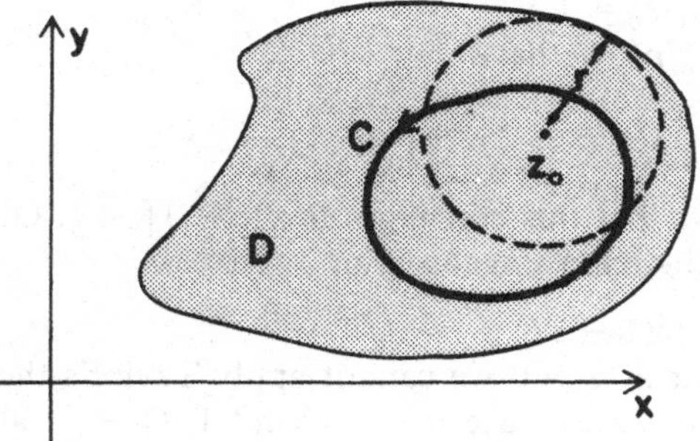

Fig. 5.15: The circle of convergence of the Taylor-series expansion about z_0 of a function $f(z)$ analytic in D.

The result is summarized in the following theorem.

Taylor's theorem: If $f(z)$ is analytic in a domain D and z_0 is any point of D, let r be the radius of the largest circle with center z_0 lying wholly in D (see Fig. 5.15). Then there exists a power series

$$\sum_{n=0}^{\infty} a_n(z - z_0)^n,$$

which converges to $f(z)$ for $|z - z_0| < r$, and

$$a_n = \frac{f^{(n)}(z_0)}{n!} = \frac{1}{2\pi i}\oint_C \frac{f(\xi)}{(\xi - z_0)^{n+1}}\,d\xi, \tag{5.30}$$

where C is any closed path in D surrounding z_0.

The equivalence between the analyticity of a function and the convergence of a Taylor-series expansion is now complete. If $f(z)$ is analytic in a domain of rather general shape, we cannot expect to represent $f(z)$ throughout the entire domain using only a single power series, and since the series converges within a circle, it may be necessary to choose several (perhaps infinitely many) points z_0 about which to expand the function. But it can always be done, and Taylor's theorem provides the means for generating the series.

In Examples 1–3 find the Taylor-series expansion about the designated point z_0 for each of the functions $f(z)$.

Example 1. $f(z) = e^z$, $z_0 = 0$.
Since $f^{(n)}(0) = 1$, we have

$$e^z = 1 + z + \frac{z^2}{2!} + \frac{z^3}{3!} + \cdots,$$

which converges for all finite z.

Example 2. $f(z) = 1/z$, $z_0 = 1$.

Since $f^{(n)}(1) = (-1)^n n!$, we have

$$\frac{1}{z} = 1 - (z-1) + (z-1)^2 - (z-1)^3 + \cdots$$

and the ratio test shows that this converges for $|z-1| < 1$. Obviously, the region does not include the singularity at $z = 0$.

Example 3. $f(z) = 1/[z^2(1-z)]$, $z_0 = 0$.

Since $f(z)$ is singular at $z = 0$ we cannot apply Taylor's theorem to $f(z)$ in its entirety, but we can use it to expand $1/(1-z)$ about the origin. For $|z| < 1$

$$\frac{1}{1-z} = 1 + z + z^2 + z^3 + \cdots$$

and hence

$$f(z) = \frac{1}{z^2}(1 + z + z^2 + z^3 + \cdots)$$
$$= \frac{1}{z^2} + \frac{1}{z} + 1 + z + \cdots$$

for $0 < |z| < 1$.

In this last example the designated point of expansion lies outside the domain of analyticity of the function and illustrates the possibility of a power series that is more general than those considered until now.

Laurent series

The series considered so far involve only nonnegative powers of the variable and converge in circular regions where they represent analytic functions. As demonstrated by the last example, these series are not always convenient or even adequate for the functions of concern to us, and it is necessary to enlarge their scope by allowing negative powers as well.

Consider the series

$$\sum_{n=1}^{\infty} b_n (z-z_0)^{-n} = \frac{b_1}{z-z_0} + \frac{b_2}{(z-z_0)^2} + \cdots . \tag{5.31}$$

The substitution $z_1 = 1/(z-z_0)$ converts this to

$$\sum_{n=1}^{\infty} b_n z_1^n,$$

which converges inside some circle $|z_1| = r^*$, where it represents an analytic function. It follows immediately that the series (5.31) converges for $|z - z_0| > 1/r^* = r_1$, that is, outside a circle of radius r_1 centered on $z = z_0$, and there represents an analytic function. If $r_1 = 0$ the series converges for all z except $z = z_0$, where the function is obviously singular, whereas if $r_1 = \infty$ the series converges only at infinity.

If we add to (5.31) the Taylor-series (5.27) converging for (say) $|z - z_0| < r_2$, we obtain the sum

$$\sum_{n=0}^{\infty} a_n(z - z_0)^n + \sum_{n=1}^{\infty} b_n(z - z_0)^{-n}.$$

Assuming $r_1 < r_2$, this converges in the annular domain $r_1 < |z - z_0| < r_2$ where it represents an analytic function $f(z)$, and thus

$$f(z) = \sum_{n=-\infty}^{\infty} a_n(z - z_0)^n, \tag{5.32}$$

where we have written $a_n = b_{-n}$ for $n < 0$. Of course, if $r_1 > r_2$ there is no value of z for which both series comprising (5.32) converge and $f(z)$ is nowhere analytic. The series (5.32) is called a *Laurent series*, after the French mathematician Pierre Alphonse Laurent (1813–54), and constitutes an expansion of the function $f(z)$ about the point $z = z_0$.

Laurent's theorem: If $f(z)$ is analytic on two concentric circles C_1 and C_2 with center z_0 and in the (open) annulus between them, there exists a power series

$$\sum_{n=-\infty}^{\infty} a_n(z - z_0)^n = \sum_{n=0}^{\infty} a_n(z - z_0)^n + \sum_{n=1}^{\infty} b_n(z - z_0)^{-n},$$

which converges to $f(z)$ for all z in the annulus, and

$$a_n = \frac{1}{2\pi i}\oint_C \frac{f(\xi)}{(\xi - z_0)^{n+1}}\,d\xi, \tag{5.33}$$

where C is any closed path lying in the annulus and enclosing the circle C_1 (see Fig. 5.16).

Proof: If z is any point in the annulus, it follows from Cauchy's integral formulas (5.25) that

$$f(z) = \frac{1}{2\pi i}\oint_{C_2} \frac{f(\xi)}{\xi - z}\,d\xi - \frac{1}{2\pi i}\oint_{C_1} \frac{f(\xi)}{\xi - z}\,d\xi. \tag{5.34}$$

Since z lies inside C_2, $|z - z_0| < |\xi - z_0|$, and the first integral is similar to that considered in the proof of Taylor's theorem. Hence

$$\frac{1}{2\pi i}\oint_{C_2} \frac{f(\xi)}{\xi - z}\,d\xi = \sum_{n=0}^{\infty} a_n(z - z_0)^n$$

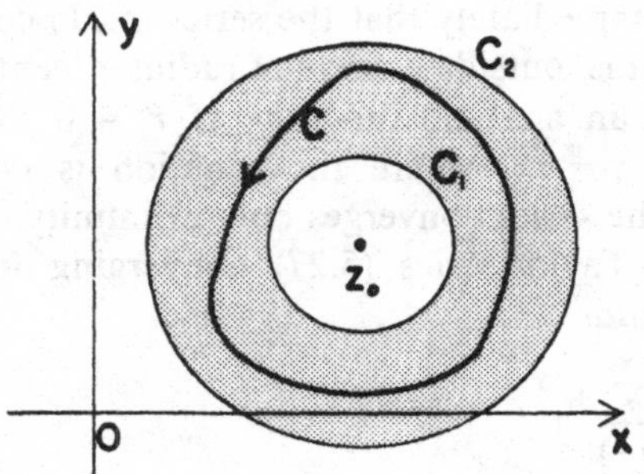

Fig. 5.16: Laurent's theorem.

with

$$a_n = \frac{1}{2\pi i}\oint_{C_2} \frac{f(\xi)}{(\xi - z_0)^{n+1}}\, d\xi, \qquad n \geq 0.$$

The series converges for all z in the annulus, and because $f(\xi)/(\xi - z_0)^{n+1}$ is analytic for ξ in this region, C_2 can be deformed into the path C shown in Fig. 5.16.

For the second integral in (5.34), $|z - z_0| > |\xi - z_0|$, and we therefore write

$$\frac{1}{\xi - z} = -\frac{1}{z - z_0}\left(1 - \frac{\xi - z_0}{z - z_0}\right)^{-1},$$

which can be expanded as

$$\frac{1}{\xi - z} = -\frac{1}{z - z_0}\left\{1 + \frac{\xi - z_0}{z - z_0} + \left(\frac{\xi - z_0}{z - z_0}\right)^2 + \cdots\right\}.$$

The series is uniformly convergent for all z in the annulus, and when it is used to replace $1/(\xi - z)$ in the second integral in (5.34), the integration can be carried out term by term, giving

$$-\frac{1}{2\pi i}\oint_{C_1} \frac{f(\xi)}{\xi - z}\, d\xi = \frac{1}{2\pi i}\left\{\frac{1}{z - z_0}\oint_{C_1} f(\xi)\, d\xi \right.$$

$$\left. + \frac{1}{(z - z_0)^2}\oint_{C_1} (\xi - z_0) f(\xi)\, d\xi + \cdots\right\}$$

$$= \sum_{n=1}^{\infty} b_n (z - z_0)^{-n}$$

with

$$b_n = \frac{1}{2\pi i}\oint_{C_1} (\xi - z_0)^{n-1} f(\xi)\, d\xi.$$

If $a_n = b_{-n}$ for $n < 0$, then

$$a_n = \frac{1}{2\pi i}\oint_{C_1} \frac{f(\xi)}{(\xi - z_0)^{n+1}}\, d\xi, \qquad n < 0$$

and since C_1 can also be deformed into the path C, the proof is now complete. □

Starting with the annulus specified in the theorem, we can progressively decrease the radius of C_1 and increase the radius of C_2 until each of the circles reaches a point at which $f(z)$ is singular, and the series will converge and represent $f(z)$ at all points of the resulting annulus. We also note that the coefficients a_n have the same mathematical expression for all n. For $n \geq 0$ we cannot use the general integral formula (5.26) to conclude that $a_n = f^{(n)}(z_0)/n!$, since f is not necessarily analytic everywhere inside C_2. Similarily, for $n < 0$ (when $a_n = b_{-n}$) the coefficients are not, in general, zero. But if f *is* analytic at all points inside C_2, the coefficients b_n are zero from Cauchy's theorem (5.21), and the Laurent series reduces to the Taylor-series expansion of $f(z)$ about z_0, as it should.

If f is singular at one or more points inside C, the Laurent series will contain at least one and perhaps an infinity of negative powers. In this case the evaluation of the integral expression for a_n is not trivial, but as the following examples show, it is generally possible to avoid using (5.33) and to compute the coefficients rather easily. The examples also show that in different annuli with the same center z_0 a function may have different Laurent expansions, but in a given annulus the Laurent expansion of a function is unique.

In Examples 1–3 find the Laurent series expansion(s) about the designated point z_0 for each of the functions $f(z)$.

Example 1. $f(z) = 1/(z-1)$, $z_0 = 0$.

Since $f(z)$ is singular at $z = 1$, there are two possible expansions: one for $|z| < 1$ and the other for $|z| > 1$. In the circular domain $|z| < 1$ the function is analytic and the Laurent expansion is simply the Taylor series. Because $f^{(n)}(0) = -n!$, we have

$$f(z) = -1 - z - z^2 - \cdots, \qquad |z| < 1.$$

For the region $|z| > 1$ the key step is to write

$$\frac{1}{z-1} = \frac{1}{z}\cdot\frac{1}{1 - 1/z}.$$

The factor $1/z$ is a power of $z - z_0$ with $z_0 = 0$ and is therefore in the required form already, so it is only necessary to expand the other factor in powers of z. From the Taylor expansion of $1/(1 - q)$ for $|q| < 1$, the substitution $q = 1/z$ gives

$$\frac{1}{1 - 1/z} = 1 + \frac{1}{z} + \frac{1}{z^2} + \cdots, \qquad |z| > 1,$$

and hence

$$f(z) = \frac{1}{z} + \frac{1}{z^2} + \frac{1}{z^3} + \cdots,$$

which is the desired Laurent expansion for $|z| > 1$. Note that if we had asked for the expansion(s) about $z_0 = 1$, $f(z)$ is in the required form as given, and no further expansion is possible.

Example 2. $f(z) = 1/[z(z - 1)]$, $z_0 = 0$.

Since $f(z)$ differs from the function discussed above only by $1/z$, which is analytic for $z \neq 0$ and already expressed as a power of z, it follows immediately that

$$\begin{aligned} f(z) &= -\frac{1}{z} - 1 - z - \cdots, \qquad 0 < |z| < 1 \\ &= \frac{1}{z^2} + \frac{1}{z^3} + \frac{1}{z^4} + \cdots, \qquad |z| > 1. \end{aligned}$$

Similarly, for $f(z) = 1/[(z - 1)(z - 2)]$ with $z_0 = 1$,

$$\begin{aligned} f(z) &= -\frac{1}{z - 1} - 1 - (z - 1) - \cdots, \qquad 0 < |z - 1| < 1 \\ &= \frac{1}{(z - 1)^2} + \frac{1}{(z - 1)^3} + \frac{1}{(z - 1)^4} + \cdots \qquad |z - 1| > 1. \end{aligned}$$

Example 3. $f(z) = (\sin z)/z^2$, $z_0 = 0$.

The numerator is an analytic function for all (finite) z and its Taylor series is

$$\sin z = z - \frac{z^3}{3!} + \frac{z^5}{5!} - \cdots, \qquad |z| < \infty.$$

Hence

$$f(z) = \frac{1}{z} - \frac{z}{3!} + \frac{z^3}{5!} - \cdots, \qquad 0 < |z| < \infty.$$

As evidenced by these examples, a Laurent expansion can usually be obtained in a rather simple fashion using Taylor series and a little algebra.

Isolated singularities

If $f(z)$ is analytic in a neighborhood of z_0 except at z_0 itself, $f(z)$ is said to be analytic in a *deleted neighborhood* of z_0 and the point z_0 is called an *isolated singularity*. Not all singularities of a function are isolated (a branch point is an example of one that is not), but isolated singularities are the most important ones for practical purposes.

A deleted neighborhood $0 < |z - z_0| < r_2$ is a special case of the annular domain to which Laurent's theorem is applicable, and in such a neighborhood the function $f(z)$ can be expanded in a Laurent series:

$$f(z) = \sum_{n=-\infty}^{\infty} a_n(z - z_0)^n$$
$$= \sum_{n=0}^{\infty} a_n(z - z_0)^n + \sum_{n=1}^{\infty} \frac{b_n}{(z - z_0)^n}.$$

The positive-power (or Taylor-series) portion converges for $|z - z_0| < r_2$ and represents a function that is analytic at $z = z_0$, whereas the negative-power portion converges for all z except at $z = z_0$. The latter, namely

$$\sum_{n=-1}^{-\infty} a_n(z - z_0)^n = \sum_{n=1}^{\infty} \frac{b_n}{(z - z_0)^n},$$

is called the *principal part* of $f(z)$ at $z = z_0$, and its form leads to a classification of the isolated singularity z_0.

There are three possible cases.

(i) *No negative powers of $z - z_0$ appear*, that is, $a_n = 0$ for $n < 0$. The principal part is then zero and $f(z)$ is analytic at $z = z_0$. In spite of any impressions to the contrary, $z = z_0$ is not in fact a singularity. As an example, $z = 0$ is *not* a singularity of $f(z) = (\sin z)/z$ since its Laurent expansion for $0 < |z| < \infty$ is

$$\frac{\sin z}{z} = 1 - \frac{z^2}{3!} + \frac{z^4}{5!} - \cdots .$$

(ii) *A finite number of negative powers of $z - z_0$ appear*, that is,

$$f(z) = \sum_{n=0}^{\infty} a_n(z - z_0)^n + \frac{a_{-1}}{z - z_0}$$
$$+ \frac{a_{-2}}{(z - z_0)^2} + \cdots + \frac{a_{-N}}{(z - z_0)^N} \tag{5.35}$$

for $N \geq 1$ with $a_{-N} \neq 0$. The function $f(z)$ is said to have a pole of order N at $z = z_0$, and if $N = 1$ the pole is sometimes referred to as a *simple pole*. Conversely, if $z = z_0$ is a pole of order N, $f(z)$ has the representation shown in Eq. (5.35) in a deleted neighborhood of z_0. A function whose only singularities in the finite part of the plane are poles is called a *meromorphic function*, and a typical example is a rational function, for example,

$$f(z) = \frac{z+1}{(z^2+1)(z-1)^2},$$

which has first-order (simple) poles at $z = \pm i$ and a second-order pole at $z = 1$.

(iii) *An infinity of negative powers of* $(z - z_0)$ *appear*. In this case $f(z)$ is said to have an *essential singularity* at $z = z_0$. An illustration is provided by the function $e^{1/z}$ whose Laurent expansion is

$$e^{1/z} = 1 + \frac{1}{z} + \frac{1}{2!}\frac{1}{z^2} + \frac{1}{3!}\frac{1}{z^3} + \cdots$$

for $|z| > 0$. Thus, $z = 0$ is an essential singularity. For practical purposes such singularities are of no concern. The functional behavior in the vicinity of an essential singularity can be very complicated indeed, and the isolated singularities of interest to us are poles.

To determine rigorously the nature of a singularity it is necessary to compute the Laurent expansion of the function in a deleted neighborhood of the point in question, with emphasis on the principal part. As an example, consider the singularity of

$$f(z) = \frac{1+z}{z^3(1-z)}$$

at $z = 0$ where the function is clearly infinite. The factor $1/z^3$ is already in the required form, and since $1/(1-z)$ is analytic at $z = 0$, it can be expanded in a Taylor series:

$$\frac{1}{1-z} = 1 + z + z^2 + z^3 + \cdots, \qquad |z| < 1.$$

Hence

$$\frac{1+z}{1-z} = 1 + 2z + 2z^2 + 2z^3 + \cdots$$

and

$$f(z) = \frac{1}{z^3} + \frac{2}{z^2} + \frac{2}{z} + 2 + \cdots$$

for $0 < |z| < 1$. The first three terms constitute the principal part, which shows that $z = 0$ is a third-order pole. This is otherwise evident from that fact that

$$\frac{1+z}{z^3(1-z)} \sim \frac{1}{z^3} \quad \text{as } z \to 0.$$

If a Laurent expansion of a function is given, care is necessary to ensure that the expansion converges in a deleted neighborhood of a point before drawing a conclusion about the singularity. Thus, if

$$f(z) = -\frac{1}{z^3} - \frac{2}{z^4} - \frac{2}{z^5} - \frac{2}{z^6} - \cdots,$$

it is tempting to conclude that $z = 0$ is an essential singularity, but the ratio test shows that the expansion converges for $|z| > 1$ and not near $z = 0$. Indeed, with a little juggling the expansion can be written as

$$\begin{aligned}(1+z)\left\{-\frac{1}{z^4} - \frac{1}{z^5} - \frac{1}{z^6} - \cdots\right\} &= -\frac{1}{z^4}(1+z) \\ &\quad \times\left\{1 + \frac{1}{z} + \frac{1}{z^2} + \cdots\right\} \\ &= -\frac{1}{z^4}(1+z)\frac{1}{1-1/z},\end{aligned}$$

which is identical to the function discussed above. The singularity at $z = 0$ is therefore a third-order pole.

Exercises

For all except Exercises 2 and 5 it is sufficient to determine only those coefficients necessary to make the form of the expansion evident.

1. Find the Taylor-series expansions of the following functions about the designated point z_0 and state the region of convergence.
   *(i) $\cosh z$, $z_0 = 0$. (ii) $\cos z$, $z_0 = \pi/4$.
   (iii) $\left(\frac{1+z}{1-z}\right)^2$, $z_0 = 0$. (iv) $\frac{z-1}{z^3}$, $z_0 = 1$.
2. Obtain the Taylor-series representation

$$z\cosh z^2 = z + \sum_{n=1}^{\infty} \frac{1}{(2n)!} z^{4n+1}, \qquad |z| < \infty.$$

3. Find all of the Taylor- and Laurent-series expansions of

$$\frac{z^2 - 1}{(z+2)(z+3)}$$

about the origin.

4. Find the Laurent-series expansions of the following functions in the domain indicated.

*(i) $\dfrac{z}{z^2 - 1}$, $|z| > 1$. (ii) $\dfrac{1}{(z-1)(z-2)}$, $|z| < 1$.

(iii) $\dfrac{1}{(z-1)(z-2)}$, $|z| > 2$.

(iv) $\dfrac{e^z \cos z}{z^3}$, $0 < |z| < \infty$.

5. Obtain the Laurent-series representation

$$\frac{\sinh z}{z^2} = \frac{1}{z} + \sum_{n=1}^{\infty} \frac{1}{(2n+1)!} z^{2n-1}, \qquad 0 < |z| < \infty.$$

6. Find the Laurent-series expansions of the following functions in a deleted neighborhood of the origin.

(i) $\dfrac{e^z - 1}{z}$. (ii) $\dfrac{z - \sin z}{z^3}$.

(ii) $\dfrac{1}{z^2(z-3)}$. (iv) $\dfrac{1}{\sin z}$.

7. For each of the following functions, find the principal part at the origin and classify the singularity there.

*(i) $\dfrac{z^2 + 3z + 1}{z^4}$. (ii) $\dfrac{z^2 - 2}{z(z+1)}$.

8. Find the principal part at the designated point and classify the singularity.

(i) $\dfrac{e^z \sin z}{(z-1)^2}$ at $z = 1$. (ii) $\dfrac{1}{z^3(z^3 + z + 1)}$ at $z = 0$.

9. Find the principal part at the designated point and classify the singularity.

(i) $\dfrac{1 - e^{-(z-1)}}{z(z-1)^4}$ at $z = 1$.

(ii) $\dfrac{z}{(z+1)^2(z^3+2)}$ at $z = -1$.

10. Classify the singularities $z = 0$ and $z = 1$ of the function

$$f(z) = \frac{z-2}{z^3} \sin \frac{1}{1-z}.$$

5.6 Residues and the residue theorem

Suppose $f(z)$ is analytic throughout a domain D except for an isolated singularity at z_0. If C is a closed contour surrounding z_0 and lying wholly in D, the integral

$$\oint_C f(z)\,dz$$

will not in general be zero because of the singularity, but since its value is the same for all paths C surrounding z_0, it is a property of the singularity alone. This value, divided by $2\pi i$, is the *residue* of $f(z)$ at z_0, and will be denoted by $\text{Res}\{f(z), z_0\}$. Thus

$$\text{Res}\{f(z), z_0\} = \frac{1}{2\pi i}\oint_C f(z)\,dz, \tag{5.36}$$

where C is any closed path surrounding z_0, with $f(z)$ analytic inside and on C except at z_0.

By inserting the Laurent-series representation (5.32) for $f(z)$ into (5.36), we have

$$\text{Res}\{f(z), z_0\} = \sum_{n=-\infty}^{\infty} a_n \frac{1}{2\pi i}\oint_C (z - z_0)^n\,dz,$$

and if C is deformed into a circle of unit radius with center at z_0, the integral can be evaluated using Eqs. (5.16) and (5.17). It follows that

$$\frac{1}{2\pi i}\oint_C (z - z_0)^n\,dz = \begin{cases} 0 & \text{if } n \neq -1, \\ 1 & \text{if } n = -1, \end{cases}$$

and hence

$$\text{Res}\{f(z), z_0\} = a_{-1}, \tag{5.37}$$

where a_{-1} is the coefficient of $1/(z - z_0)$ in the Laurent expansion of $f(z)$ in the vicinity of z_0. By virtue of the definition (5.36) of a residue, (5.37) is also obtained immediately by putting $n = -1$ in Eq. (5.33).

The extension to the case when there are several isolated singularities inside C is rather obvious and is covered by the following theorem.

Cauchy's residue theorem

If $f(z)$ is analytic inside and on a closed contour C except for a finite number of isolated singularities $z_1, z_2, \ldots, z_m$ inside C, then

$$\oint_C f(z)\,dz = 2\pi i \sum_{j=1}^{m} \text{Res}\{f(z), z_j\}. \tag{5.38}$$

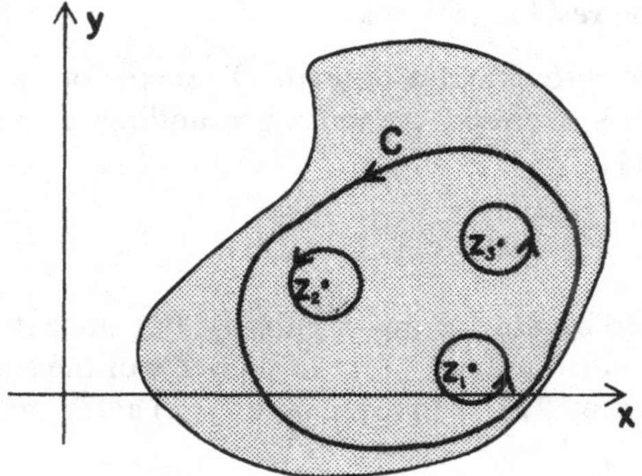

Fig. 5.17: Cauchy's residue theorem.

Proof: Surround each of the singularities z_j with a circle C_j whose radius is small enough to have the m circles and C separated from one another (see Fig. 5.17). Since the circles plus C enclose a region in which $f(z)$ is analytic,

$$\oint_C f(z)\,dz - \oint_{C_1} f(z)\,dz - \cdots - \oint_{C_m} f(z)\,dz = 0$$

by Cauchy's integral theorem, and hence

$$\oint_C f(z)\,dz = \sum_{j=1}^{m} \oint_{C_j} f(z)\,dz = 2\pi i \sum_{j=1}^{m} \operatorname{Res}\{f(z), z_j\}. \quad \square$$

The proof is only a simple exercise in path deformation, but the theorem itself is fundamental to integration in the complex plane. If it were possible to remember (and know how to use) only one theorem dealing with functions of a complex variable, this would be the one to choose. It includes Cauchy's integral theorem as a special case, as well as Cauchy's integral and general integral formulas, and reduces the evaluation of a closed contour integral to a residue calculation.

Residue computation

A residue is a number, in general complex, equal to the coefficient a_{-1} in the principal part or negative-power portion of the Laurent-series expansion of the function in a deleted neighborhood of the singularity. If the expansion is available, the residue is obviously known, but it is not necessary to obtain the complete expansion in order to compute the residue. As we shall now show, there is a simpler approach when the singularity is a pole.

(i) First-order pole

If $z = z_0$ is a first-order (simple) pole of $f(z)$, the Laurent-series expansion of $f(z)$ in the immediate vicinity is

$$f(z) = \phi(z) + \frac{a_{-1}}{z - z_0},$$

where

$$\phi(z) = \sum_{n=0}^{\infty} a_n(z - z_0)^n$$

is the positive-power or Taylor-series portion, representing a function that is analytic at $z = z_0$. Let

$$g(z) = (z - z_0)f(z) \tag{5.39}$$

so that

$$g(z) = a_{-1} + (z - z_0)\phi(z).$$

Since $\phi(z)$ is finite at $z = z_0$, it follows that $a_{-1} = g(z_0)$, and the required formula is therefore

$$\text{Res}\{f(z), z_0\} = \lim_{z \to z_0} \{(z - z_0)f(z)\}. \tag{5.40}$$

As an example, if

$$f(z) = \frac{z + 1}{z^2(z - 1)}e^z,$$

$$\text{Res}\left\{\frac{z + 1}{z^2(z - 1)}e^z, 1\right\} = \lim_{z \to 1}\left\{\frac{z + 1}{z^2}e^z\right\} = 2e.$$

There is another formula which is sometimes advantageous. Let

$$f(z) = \frac{P(z)}{Q(z)},$$

where $P(z)$ is analytic at $z = z_0$. If z_0 is a simple pole of $f(z)$, then $[Q(z)]/(z - z_0)$ is analytic here and $Q(z)$ can be expanded in a Taylor series of the form

$$Q(z) = (z - z_0)Q'(z_0) + \frac{1}{2!}(z - z_0)^2 Q''(z_0) + \cdots.$$

Hence, from (5.40),

$$\text{Res}\{f(z), z_0\} = \lim_{z \to z_0}\left\{(z - z_0)\frac{P(z)}{Q(z)}\right\} = \frac{P(z_0)}{Q'(z_0)}. \tag{5.41}$$

As an example, consider

$$f(z) = \frac{1}{z^4 + 1}.$$

This has simple poles at $e^{\pm i\pi/4}$ and $-e^{\pm i\pi/4}$, and if z_0 is any one of these,

$$\text{Res}\left\{\frac{1}{z^4 + 1}, z_0\right\} = \frac{1}{4z_0^3} = -\frac{z_0}{4}$$

since $z_0^4 = -1$.

(ii) Nth-order pole

If $f(z)$ has a pole of order N at $z = z_0$, the corresponding Laurent series is

$$f(z) = \phi(z) + \frac{a_{-1}}{z - z_0} + \frac{a_{-2}}{(z - z_0)^2} + \cdots + \frac{a_{-N}}{(z - z_0)^N},$$

where $a_{-N} \neq 0$ and $\phi(z)$ is the positive-power portion as before. Let

$$g(z) = (z - z_0)^N f(z) \tag{5.42}$$

so that

$$g(z) = a_{-N} - a_{-N+1}(z - z_0) + \cdots + a_{-1}(z - z_0)^{N-1} + (z - z_0)^N \phi(z).$$

Clearly,

$$a_{-1} = \frac{1}{(N-1)!} g^{(N-1)}(z_0),$$

and thus

$$\text{Res}\{f(z), z_0\} = \lim_{z \to z_0} \left\{\frac{1}{(N-1)!} \frac{d^{N-1}}{dz^{N-1}}\left[(z - z_0)^N f(z)\right]\right\} \tag{5.43}$$

for $N \geq 1$. An example is provided by the third-order pole $z = 1$ of the function

$$f(z) = \frac{z^2 + 2}{z(z - 1)^3},$$

and since

$$(z - 1)^3 f(z) = z + 2/z,$$

we have

$$\operatorname{Res}\left\{\frac{z^2+2}{z(z-1)^3}, 1\right\} = \lim_{z\to 1}\left\{\frac{1}{2!}\frac{4}{z^3}\right\} = 2.$$

For a single pole inside the contour C, the residue theorem in conjunction with the expression (5.40) or (5.43) for the residue is equivalent to Cauchy's integral or general integral formula. If, for example, z_0 is a simple pole, the function $g(z)$ in Eq. (5.39) is identical to that in the integral formula (5.25), and the residue is, of course, $g(z_0)$. Similarly, if z_0 is a pole of order $N = n + 1$, the function $g(z)$ defined in Eq. (5.42) is the same as that in the general integral formula (5.26), and the value of the integral is in accordance with the expression (5.43) for the residue. Nevertheless, the residue theorem is more general in that the isolated singularities are not limited to poles (though this is the only case we consider), and there can be a finite number of singularities inside C.

The formulas (5.40) and (5.43) for the residues should seem familiar to the reader. A residue calculation is analogous to the determination of a coefficient in the partial-fraction expansion of a rational function, and the process can be carried out using the finger method described in Section 2.9. But there are differences. As noted above, the function $f(z)$ is not restricted to rational functions, and since we are only concerned with a single coefficient for each of the singularities inside C, a partial-fraction expansion is unnecessary.

Integrals of trigonometric functions

The residue theorem can be used to evaluate a class of definite integrals involving trigonometric functions, and though these can also be evaluated by other means, the theorem provides a simple and elegant approach.

The integrals are of the type

$$I = \int_0^{2\pi} F(\cos\theta, \sin\theta)\, d\theta, \tag{5.44}$$

where the function F is a ratio of polynomials in $\cos\theta$ and $\sin\theta$ that remains finite for $0 \le \theta < 2\pi$. If we regard θ as the argument of z on the unit circle, then $e^{i\theta} = z$,

$$\cos\theta = \frac{1}{2}(e^{i\theta} + e^{-i\theta}) = \frac{1}{2}\left(z + \frac{1}{z}\right),$$

$$\sin\theta = \frac{1}{2i}(e^{i\theta} - e^{-i\theta}) = \frac{1}{2i}\left(z - \frac{1}{z}\right),$$

and the integral becomes a rational function of z, say $f(z)$. Since

$dz = ie^{i\theta}\,d\theta$, we have $d\theta = dz/iz$ and thus

$$I = \oint_C f(z)\frac{dz}{iz}, \tag{5.45}$$

where the contour C is the unit circle $|z| = 1$. This can be evaluated in terms of the residues at the poles inside C.

Example. Evaluate

$$I = \oint_0^{2\pi} \frac{d\theta}{1 - 2a\cos\theta + a^2},$$

where a is real with $0 \le a^2 < 1$.

From the preceding expression for $\cos\theta$ we obtain

$$I = \oint_C \frac{dz/(iz)}{1 - a(z + 1/z) + a^2} = \frac{i}{a}\oint_C \frac{dz}{(z-a)(z - 1/a)}.$$

The integrand has simple poles at a, $1/a$ and only the former lies inside the unit circle. Since

$$\operatorname{Res}\left\{\frac{1}{(z-a)(z-1/a)}, a\right\} = -\frac{a}{1-a^2},$$

the residue theorem gives

$$I = 2\pi i \frac{i}{a}\left(-\frac{a}{1-a^2}\right) = \frac{2\pi}{1-a^2}.$$

If $a^2 > 1$, the integral differs only in sign, and if $a^2 = 1$, the integral diverges. We also note that because $\cos\theta$ is an even function about $\theta = \pi$ (i.e., $\cos(\pi - \theta) = \cos(\pi + \theta)$), the integral from 0 to π can be evaluated in this manner and is simply $I/2$.

Exercises

1. Find the residues of the following at $z = 0$.

   *(i) $\dfrac{1}{z(z-1)}$. (ii) $\dfrac{3z-4}{z^2-2z}$.

   (iii) $\dfrac{1}{z^3(z+1)}$. (iv) $\dfrac{1}{\sin z}$.

2. Find the residues of the following at the designated point.

   (i) $\dfrac{1}{z^3+1}$, $z = -1$. (ii) $\dfrac{e^{az}}{(z+1)^2}$, $z = -1$.

   (iii) $\dfrac{z+1}{z^4+2z^3}$, $z = 0$. (iv) $\tan z$, $z = \pi/2$.

3. For the following $f(z)$, evaluate $\oint_C f(z)\,dz$ where C is $|z| = 1$.

*(i) $\dfrac{z+1}{z(z^2+4)}$. (ii) $\dfrac{z}{2z^2+1}$.

(iii) $\dfrac{z^4+z^2+1}{(2z+1)(z^2+2)}$. (iv) $\dfrac{z}{8z^3+1}$.

4. For the following $f(z)$, evaluate $\oint_C f(z)\,dz$ where C is $|z| = 1$.

(i) $\dfrac{1}{\sinh 2z}$. (ii) $\dfrac{\coth z}{2z-i}$.

(iii) $\dfrac{1}{1-\cos z}$. (iv) $\dfrac{1}{z^2 \sin z}$.

5. For the following $f(z)$, evaluate $\oint_C f(z)\,dz$ where C is as shown.

(i) $\dfrac{1}{z^3(z+4)}$, $|z+2| = 3$.

(ii) $\dfrac{3z^2+2}{(z-1)(z^2+9)}$, $|z-2| = 2$.

(iii) $\dfrac{\sin z}{z^2+1}$, $|z| = 2$.

(iv) $\dfrac{3z^2-6z+1}{z^3-3z^2+z-3}$, $|z+1| = 2$.

6. Using the residue theorem, evaluate:

*(i) $\displaystyle\int_0^{2\pi} \sin^4\theta\,d\theta$. (ii) $\displaystyle\int_0^{2\pi} \frac{d\theta}{2+\cos\theta}$.

(iii) $\displaystyle\int_0^{2\pi} \frac{d\theta}{2-\cos\theta}$. (iv) $\displaystyle\int_0^{2\pi} \frac{\cos\theta}{3+\cos\theta}\,d\theta$.

7. Evaluate:

(i) $\displaystyle\int_0^{2\pi} \frac{d\theta}{3+4\cos 2\theta}$. (ii) $\displaystyle\int_0^{2\pi} \frac{\cos\theta}{13-12\cos 2\theta}\,d\theta$.

(iii) $\displaystyle\int_0^{2\pi} \frac{\sin^2\theta}{5-4\cos\theta}\,d\theta$. (iv) $\displaystyle\int_0^{2\pi} \frac{\cos^2\theta}{26-10\cos 2\theta}\,d\theta$.

8. If a and b are real, evaluate:

(i) $\displaystyle\int_0^{2\pi} \frac{\cos\theta}{1-2a\cos\theta+a^2}\,d\theta$, $a^2 < 1$.

(ii) $\displaystyle\int_0^{2\pi} \frac{\sin^2\theta}{a+b\cos\theta}\,d\theta$, $a > b > 0$.

9. Show that on the unit circle

$$\cos n\theta = \frac{1}{2z^n}(z^{2n}+1), \qquad \sin n\theta = \frac{1}{2iz^n}(z^{2n}-1)$$

for any integer n.

10. If a is real with $a^2 < 1$, show that:

(i) $$\int_0^{\pi} \frac{\cos 2\theta}{1 - 2a\cos\theta + a^2}\, d\theta = \frac{\pi a^2}{1 - a^2}.$$

(ii) $$\int_0^{\pi} \frac{\cos^2 3\theta}{1 - 2a\cos 2\theta + a^2}\, d\theta = \frac{\pi}{2}\left(1 + \frac{a^2}{1 - a}\right).$$

5.7 Closing the contour

Cauchy's residue theorem is a powerful tool for evaluating a class of integrals that arise in the solution of physical problems, but since the theorem is valid only for closed contours, it is not immediately evident how the theorem can be applied. In the previous section we showed that integrals of trigonometric functions over a 2π range could be interpreted as integrals around the unit circle, and we now consider three types of infinite integrals for which it is possible to close the contour in a particular manner. The integrals include those of interest in connection with the Fourier and inverse Laplace transforms discussed in later chapters.

We are often faced with evaluating an integral of the form

$$I = \int_{-\infty}^{\infty} f(x)\, dx, \tag{5.46}$$

where f is a real or complex function and the infinite integral is interpreted as

$$I = \lim_{R \to \infty} \int_{-R}^{R} f(x)\, dx.$$

It is assumed that the limit exists and that f has no singularities in $-\infty < x < \infty$. If $f(x)$ is an even function (i.e., $f(-x) = f(x)$), it is obvious that

$$I = 2\int_0^{\infty} f(x)\, dx, \tag{5.47}$$

whereas if $f(x)$ is odd (i.e., $f(-x) = -f(x)$), then $I = 0$.

We can regard I as a line integral along the real axis of the complex z plane, and in seeking a closed contour, of which the real axis is part, we are led to consider

$$\oint_C f(z)\, dz,$$

where C consists of the real axis from $-R$ to R and a semicircle Γ_1 of radius R in the upper half-plane (see Fig. 5.18). If $f(z)$ is analytic on C

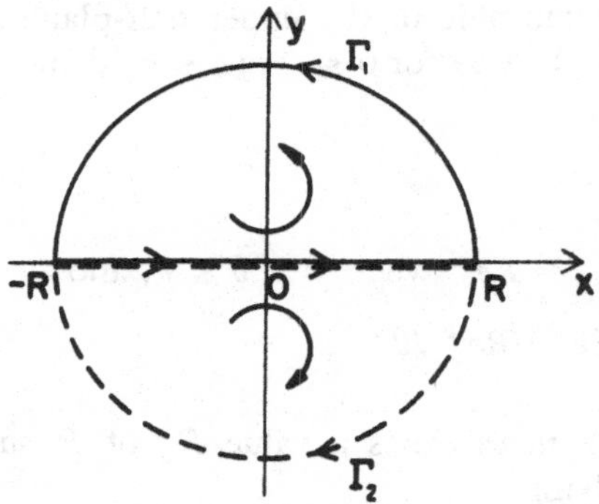

Fig. 5.18: Path closure in the z plane.

and meromorphic inside, Cauchy's residue theorem gives

$$\int_{-R}^{R} f(z)\,dz + \int_{\Gamma_1} f(z)\,dz = 2\pi i \sum_n \operatorname{Res}\{f(z), z_n\},$$

where z_n are the poles inside C, and thus

$$\int_{-R}^{R} f(z)\,dz = 2\pi i \sum_n \operatorname{Res}\{f(z), z_n\} - \int_{\Gamma_1} f(z)\,dz.$$

In the limit as $R \to \infty$ the left-hand side is I and the interior of C is the entire upper half-plane (UHP). We then have

$$I = 2\pi i \sum \operatorname{Res}\{f(z), \mathrm{UHP}\} - \lim_{R\to\infty} \int_{\Gamma_1} f(z)\,dz. \tag{5.48}$$

Similarly, if Γ_2 is a closing semicircle in the lower half-plane (LHP),

$$I = -2\pi i \sum \operatorname{Res}\{f(z), \mathrm{LHP}\} - \lim_{R\to\infty} \int_{\Gamma_2} f(z)\,dz, \tag{5.49}$$

where the minus sign occurs because the closed contour is described clockwise. Equation (5.48) or (5.49) provides a method for computing I if we can determine the limiting value of the contribution from the semicircle Γ_1 or Γ_2, and the only significant case is that in which the limiting value can be shown to be zero.

Rational functions

If $f(z)$ is a rational function, that is, the ratio of two polynomials in z, the condition under which the contour can be closed can be deduced from the following lemma.

Lemma 1: If (i) $f(z)$ is meromorphic in the upper half-plane and (ii) $zf(z) \to 0$ uniformly as $|z| = R \to \infty$ for $0 \leq \arg z \leq \pi$, then

$$\lim_{R\to\infty} \int_{\Gamma_1} f(z)\,dz = 0$$

Proof: On the semicircle Γ_1, $z = Re^{i\theta}$ where $0 \leq \theta \leq \pi$, and

$$\int_{\Gamma_1} f(z)\,dz = \int_0^{\pi} f(Re^{i\theta})iRe^{i\theta}\,d\theta.$$

In view of the condition (ii), there exists a value R_0 of R such that $|zf(z)| < \varepsilon$ for $R > R_0$, implying

$$\left|\int_{\Gamma_1} f(z)\,dz\right| \leq \int_0^{\pi} |Rf(Re^{i\theta})|\,d\theta < \varepsilon \int_0^{\pi} d\theta = \varepsilon\pi.$$

Since this can be made as small as we like by choosing R_0 sufficiently large, it follows that

$$\lim_{R\to\infty} \int_{\Gamma_1} f(z)\,dz = 0.$$

Subject to the same conditions in the lower half-plane, we also have

$$\lim_{R\to\infty} \int_{\Gamma_2} f(z)\,dz = 0. \quad \square$$

A particular example of a meromorphic function is a rational function, and condition (ii) now requires that the degree of the denominator polynomial exceed that of the numerator polynomial by *at least* 2. If this is so, for $|z|$ sufficiently large, $f(z)$ behaves like z^{-p} with $p \geq 2$, and the conditions for path closure are satisfied in both the upper and the lower half-planes. We have a *free choice* of the half-plane in which to close the contour, and the resulting expressions for I are

$$I = 2\pi i \sum \text{Res}\{f(z), \text{UHP}\}, \tag{5.50}$$

$$I = -2\pi i \sum \text{Res}\{f(z), \text{LHP}\}. \tag{5.51}$$

Naturally, both give the same answer, but if f is a complex function, a wise choice of half-plane can simplify the calculation.

Example 1. Evaluate $I = \int_{-\infty}^{\infty} dx/(x^2 + 1)$.

To set the stage for what we are about to do, it is convenient to write the integral in terms of z. This is a worthwhile step under all circumstances and is particularly desirable for some of the later examples. Thus

$$I = \int_{-\infty}^{\infty} f(z)\, dz$$

where

$$f(z) = \frac{1}{z^2+1} = \frac{1}{(z-i)(z+i)}.$$

This has simple poles at $z = i$ and $z = -i$ in the upper and lower half-planes, respectively, and since the conditions for path closure are met, we can evaluate the integral using either (5.50) or (5.51). If we choose the UHP we have

$$I = 2\pi i \operatorname{Res}\{f(z), i\} = 2\pi i \frac{1}{2i} = \pi,$$

and it is easily verified that Eq. (5.51) gives the same result. Because f is real, I is real, as it must be.

Example 2. Evaluate

$$I = \int_{-\infty}^{\infty} \frac{dx}{(x^2-2ix-4)(x^2+1)}.$$

The integral can be written as

$$I = \int_{-\infty}^{\infty} f(z)\, dz,$$

where

$$f(z) = \frac{1}{(z^2-2iz-4)(z^2+1)}$$

$$= \frac{1}{(z-i-\sqrt{3})(z-i+\sqrt{3})(z-1)(z+i)}.$$

This has simple poles at $z = i$ and $z = i \pm \sqrt{3}$ in the UHP, and $z = -i$ in the LHP. The conditions for path closure are again met, and it is now easier to choose the LHP. Thus

$$I = -2\pi i \operatorname{Res}\{f(z), -i\} = -2\pi i\left(-\tfrac{1}{7}\right)\left(-\tfrac{1}{2i}\right) = -\frac{\pi}{7},$$

which is real because the imaginary part of $f(x)$ is an odd function.

Rational functions and exponentials

The function f is now the product of a rational function and an exponential with a purely imaginary exponent, and the conditions for path closure are given by the following lemma.

Lemma 2: If $f(z) = e^{iaz}g(z)$ where (i) a is real with $a > 0$, (ii) $g(z)$ is a rational function, and (iii) the degree of the denominator polynomial exceeds that of the numerator polynomial by *at least 1*, that is, the rational function is proper, then

$$\lim_{R\to\infty} \int_{\Gamma_1} f(z)\, dz = 0.$$

This is a special case of *Jordan's lemma*, named after the French mathematician Camille Jordan (1838–1921).

Proof: On the semicircle Γ_1, $z = Re^{i\theta}$ with $0 \le \theta \le \pi$, and

$$e^{iaz} = e^{iaR\cos\theta}e^{-aR\sin\theta}.$$

If R is sufficiently large, condition (iii) implies $|g(z)| < k/R$ for some finite k, so that

$$\left|\int_{\Gamma_1} f(z)\, dz\right| \le \int_0^{\pi}\left|e^{iaz}g(z)\frac{dz}{d\theta}\right| d\theta$$

$$= \int_0^{\pi} e^{-aR\sin\theta}|Rg(Re^{i\theta})|\, d\theta$$

$$< 2k\int_0^{\pi/2} e^{-aR\sin\theta}\, d\theta.$$

As θ increases from 0 to $\pi/2$, $(\sin\theta)/\theta$ decreases from 1 to $2/\pi$, and it is easily verified that $\sin\theta \ge 2\theta/\pi$. Hence

$$\left|\int_{\Gamma_1} f(z)\, dz\right| < 2k\int_0^{\pi/2} e^{-2aR\theta/\pi}\, d\theta$$

$$= \frac{\pi k}{aR}(1 - e^{-aR}) < \frac{\pi k}{aR},$$

which tends to zero as $R \to \infty$. □

For large $|z|$ the behavior of the integrand is determined by the exponential, and it is necessary to select the half-plane in which the exponential tends to zero as $|z| \to \infty$. If $a > 0$, the half-plane required for closure of the path is the upper one. Similarly, if $a < 0$,

$$\lim_{R\to\infty} \int_{\Gamma_2} f(z)\, dz = 0$$

and the contour must be closed in the lower half-plane. No longer is there freedom of choice, and the half-plane is determined by the sign of the exponent. Thus, if $f(z) = e^{iaz}g(z)$ where $g(z)$ is a proper rational function with no poles on the real axis,

$$I = 2\pi i \sum \text{Res}\{f(z), \text{UHP}\}, \qquad \text{for } a > 0 \tag{5.52}$$

and

$$I = -2\pi i \sum \text{Res}\{f(z), \text{LHP}\}, \qquad \text{for } a < 0. \tag{5.53}$$

For $a = 0$ the integral is similar to those previously considered, and if the degree of the denominator polynomial exceeds that of the numerator polynomial by only 1, the integral diverges.

It is not worth trying to remember the half-plane required for a particular a: a little "doodle" taking only a few seconds will always give the information. With $z = x + iy$ we have

$$e^{iaz} = e^{iax}e^{-ay}$$

giving

$$|e^{iaz}| = e^{-ay}.$$

Since the exponent must be negative in the half-plane employed,

if $\quad a < 0 \quad$ we need $\quad y < 0, \quad$ which is the LHP,

and

if $\quad a > 0 \quad$ we need $\quad y > 0, \quad$ which is the UHP.

Of course, to use this effectively it is necessary to write the exponential in terms of z.

Example 1. Evaluate

$$I = \int_{-\infty}^{\infty} \frac{e^{i\omega t}}{\omega - i}\, d\omega$$

for real $t \neq 0$.

The integral can be written as

$$I = \int_{-\infty}^{\infty} f(z)\, dz,$$

where

$$f(z) = \frac{e^{izt}}{z - i}.$$

This has a simple pole at $z = i$ and the conditions for closing the contour are satisfied in the half-plane specified by a doodle. In its most compact form, suitable for the margin of the paper, the doodle is

$$e^{izt} = e^{ixt}e^{-yt};$$

want $\quad yt > 0: \quad t < 0 \Rightarrow y < 0$ (LHP), $\qquad t > 0 \Rightarrow y > 0$ (UHP).

If $t < 0$ the closure must be carried out in the LHP, and since $f(z)$ is analytic there, $I = 0$. If $t > 0$, closure in the UHP gives

$$I = 2\pi i \operatorname{Res}\{f(z), i\} = 2\pi i e^{-t}.$$

Hence

$$I = 2\pi i e^{-t} u(t),$$

where $u(t)$ is the unit-step function. We note that I is discontinuous at $t = 0$, for which the integral diverges.

Example 2. Evaluate

$$I = \int_{-\infty}^{\infty} \frac{e^{-ix}}{x^2 + 1} \, dx.$$

As before, the integral is written as

$$I = \int_{-\infty}^{\infty} f(z) \, dz,$$

where

$$f(z) = \frac{e^{-iz}}{z^2 + 1} = \frac{e^{-iz}}{(z - i)(z + i)}.$$

Since

$$e^{-iz} = e^{-ix}e^{y},$$

we require $y < 0$ (LHP), and the conditions for closing the contour are satisfied in the lower half-plane. The only singularity there is a simple pole at $z = -i$, and hence

$$I = -2\pi i \operatorname{Res}\{f(z), -i\} = -2\pi i\left(\frac{e^{-1}}{-2i}\right) = \frac{\pi}{e}.$$

By taking the real and imaginary parts of the integral, we deduce that

$$\int_{-\infty}^{\infty} \frac{\cos x}{x^2 + 1} \, dx = \operatorname{Re} I = \frac{\pi}{e}$$

while

$$\int_{-\infty}^{\infty} \frac{\sin x}{x^2 + 1} \, dx = -\operatorname{Im} I = 0.$$

This last result is obvious, since the integrand is an odd function of x.

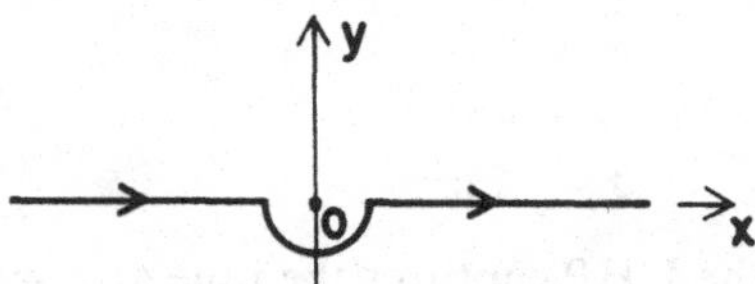

Fig. 5.19: Path indented below $z = 0$.

In the case of a real integrand which is proportional to a cosine or sine, it is generally possible (and more efficient) to consider the integral involving the corresponding exponential and to take the real or imaginary part of the result obtained. If this is not possible, for example, if $g(x)$ is complex, the trigonometric function must be written in its exponential form and each part treated separately. A special case where this is necessary is as follows:

Example 3. Evaluate

$$I = \int_{-\infty}^{\infty} \frac{\sin x}{x}\,dx.$$

From the exponential form of the sine function we obtain

$$I = \int_{-\infty}^{\infty} f(z)\,dz,$$

where

$$f(z) = \frac{1}{2i}\left(\frac{e^{iz}}{z} - \frac{e^{-iz}}{z}\right).$$

This is analytic for all finite z; in particular, $f(z)$ is not singular at $z = 0$, although the individual terms are. By carrying out the doodle, it is found that for the first exponential the closure of the contour must be in the UHP, whereas for the second exponential the LHP is required. It is therefore necessary to treat each separately, but when this is done the two integrals diverge because of the pole at $z = 0$ on the path of integration.

We can avoid the difficulty by deforming the path prior to separation. Since $f(z)$ is analytic in a neighborhood of $z = 0$, the integral around a closed contour consisting of (let us say) the real axis from $z = -\varepsilon$ to $z = \varepsilon$ and thence back to $z = -\varepsilon$, via a semicircle in the LHP, is zero by Cauchy's integral theorem. This allows us to replace the original path by the indented one shown in Fig. 5.19, and if the new path is denoted by $\oint_{-\infty}^{\infty}$, we have

$$I = \oint_{-\infty}^{\infty} f(z)\,dz = I_1 + I_2,$$

where

$$I_1 = \oint_{-\infty}^{\infty} \frac{e^{iz}}{2iz}\, dz, \qquad I_2 = -\oint_{-\infty}^{\infty} \frac{e^{-iz}}{2iz}\, dz.$$

For I_1, closure of the contour in the UHP captures the pole at $z = 0$, and thus

$$I_1 = 2\pi i \operatorname{Res}\left\{ \frac{e^{iz}}{2iz}, 0 \right\} = \pi.$$

For I_2 the closure must be in the LHP, and since there is no singularity below the path, $I_2 = 0$. Hence

$$I = I_1 + I_2 = \pi.$$

The result is independent of how the path is deformed, and if we had chosen to indent the path above the point $z = 0$, we would have obtained $I_1 = 0$ and $I_2 = \pi$, so that $I = \pi$ as before. When such a deformation of the path is needed, it must be carried out prior to the separation of the integrals, at a stage when it is justified by the analyticity of the original integrand.

Example 4. Evaluate

$$I = \int_{-i\infty}^{i\infty} \frac{e^{st}}{(s+1)^2}\, ds$$

for real t.

The integral is written as

$$I = \int_{-i\infty}^{i\infty} f(z)\, dz,$$

where

$$f(z) = \frac{e^{zt}}{(z+1)^2}$$

and we note that the path of integration is now the imaginary axis. A change of variable from z to z_1 where $z_1 = -iz$ converts I to a form similar to that of the first example, showing that the same technique for closing the contour is also applicable here, but it is unnecessary to make the transformation. The half-plane to the right or left in which the contour can be closed is again determined by the sign of the exponent, and a doodle is sufficient to specify which half-plane:

$$e^{zt} = e^{xt} e^{iyt};$$

want $xt < 0$: $t < 0 \Rightarrow x > 0$ (RHP), $\quad t > 0 \Rightarrow x < 0$ (LHP), where the abbreviations RHP and LHP refer to the right- and left-half-planes, respectively. If $t < 0$, the closure must be carried out in the RHP where $x > 0$. The resulting closed contour is described in a clockwise sense implying

$$I = -2\pi i \sum \text{Res}\{f(z), \text{RHP}\},$$

and since $f(z)$ is analytic here, $I = 0$. If $t > 0$, closure to the left gives

$$I = 2\pi i \sum \text{Res}\{f(z), \text{LHP}\}.$$

The only singularity in this region is a second-order pole at $z = -1$ whose residue is

$$\text{Res}\{f(z), -1\} = \lim_{z \to -1} \left\{ \frac{d}{dz} e^{zt} \right\} = te^{-t},$$

so that

$$I = 2\pi i t e^{-t}.$$

Hence

$$I = 2\pi i t e^{-t} u(t),$$

where $u(t)$ is again the unit-step function, and we observe that I is continuous at $t = 0$.

Multivalued functions

A more sophisticated application of the techniques described above is the use of contour integration to evaluate an integral of the form

$$\int_0^\infty x^{a-1} g(x)\, dx, \tag{5.54}$$

where $g(x)$ is a rational function and a is not an integer. To illustrate the method, consider

$$I = \int_0^\infty \frac{x^{a-1}}{x+1}\, dx, \qquad 0 < a < 1, \tag{5.55}$$

which we write as

$$I = \int_0^\infty f(z)\, dz$$

with

$$f(z) = \frac{z^{a-1}}{z+1}.$$

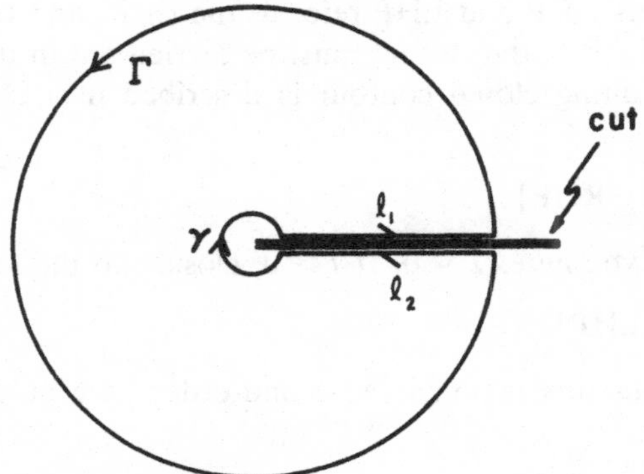

Fig. 5.20: Closed contour C in the cut plane.

This has a pole at $z = -1$ and a branch point at $z = 0$, and to ensure that $f(z)$ is uniquely defined for all z it is necessary to employ a cut plane with the branch cut extending from the origin to infinity. Since we cannot integrate *across* the cut, any closed contour that includes a circle of large radius must also include integrals along the top and bottom sides of the cut; if the cut coincides with the positive real axis, one of those integrals is I.

With the plane cut along the positive real axis or, more precisely, infinitesimally below, we define z^{a-1} as

$$z^{a-1} = |z^{a-1}|e^{i(a-1)\arg z}$$

with $0 \leq \arg z < 2\pi$, so that z^{a-1} is real when z is real and positive. Let C be the closed contour shown in Fig. 5.20, consisting of a circle Γ of large radius R, the upper and lower sides of the cut, ℓ_1 and ℓ_2 respectively, and a circle γ of small radius ε to avoid the essential singularity at $z = 0$. Inside and on C, $f(z)$ is analytic apart from the simple pole at $z = -1$ and hence, from Cauchy's residue theorem,

$$\oint_C f(z)\,dz = 2\pi i \operatorname{Res}\{f(z), -1\}$$

$$= 2\pi i e^{i(a-1)\pi}.$$

On Γ, $|zf(z)|$ behaves like R^{a-1} for large $|z| = R$, and since $a < 1$, this tends to zero as $R \to \infty$. From Lemma 1 it follows that the contribution from Γ vanishes in the limit. Similarly, on γ, $|zf(z)|$ behaves like ε^a for small $|z| = \varepsilon$, and since $a > 0$, the contribution from γ also vanishes as $\varepsilon \to 0$. We are now left with the contributions from the two sides of the cut, and as $R \to \infty$ and $\varepsilon \to 0$

$$\int_{\ell_1} f(z)\,dz \to I.$$

On the lower side $z^{a-1} = |z|^{a-1}e^{i(a-1)2\pi}$, which differs by a factor $e^{i(a-1)2\pi}$ from the value of z^{a-1} at the corresponding point on the upper side. Since ℓ_2 is also described in the opposite direction to ℓ_1,

$$\int_{\ell_2} f(z)\,dz \to -e^{i(a-1)2\pi}I,$$

and thus

$$I\{1 - e^{i(a-1)2\pi}\} = 2\pi i e^{i(a-1)\pi}$$

giving

$$I = \frac{\pi}{\sin a\pi}, \qquad 0 < a < 1.$$

The contour C shown in Fig. 5.20 can be used to evaluate the integral (5.54) when $g(x)$ is any proper rational function that is finite for all real $x > 0$.

Exercises

Evaluate the following integrals. Unless otherwise stated, a, b are any positive real numbers with $a > b$.

*1. $\displaystyle\int_{-\infty}^{\infty} \frac{dx}{(x^2+1)^2}$.

2. $\displaystyle\int_{-\infty}^{\infty} \frac{x^2\,dx}{(x^2+1)(x^2+4)}$.

3. $\displaystyle\int_{0}^{\infty} \frac{dx}{(x^2+1)^3}$.

4. $\displaystyle\int_{0}^{\infty} \frac{x^2+1}{x^4+1}\,dx$.

5. $\displaystyle\int_{-\infty}^{\infty} \frac{dx}{x^6+1}$.

6. $\displaystyle\int_{-\infty}^{\infty} \frac{x^2}{x^6+1}\,dx$.

7. $\displaystyle\int_{-\infty}^{\infty} \frac{dx}{4x^2+2x+1}$.

8. $\displaystyle\int_{-\infty}^{\infty} \frac{x\,dx}{(x^2-2x+2)^2}$.

*9. $\displaystyle\int_{-\infty}^{\infty} \frac{dx}{(x^2+a^2)(x^2+b^2)}$.

10. $\displaystyle\int_{-\infty}^{\infty} \frac{dx}{(x^2+a^2)(x^2+b^2)^2}$.

11. $\displaystyle\int_{-\infty}^{\infty} \frac{dx}{x^4+a^4}$.

12. $\displaystyle\int_{-\infty}^{\infty} \frac{x^2\,dx}{x^4+a^4}$.

13. $\displaystyle\int_{-\infty}^{\infty} \frac{e^{ix}+2e^{-ix}}{x-i}\,dx$.

14. $\displaystyle\int_{-\infty}^{\infty} \frac{\sin x}{x(x^2+1)}\,dx$.

15. $\displaystyle\int_{-\infty}^{\infty} \frac{e^{ix}}{(x^2 + a^2)(x^2 + b^2)}\,dx.$ 16. $\displaystyle\int_{-\infty}^{\infty} \frac{e^{iax}}{(x^2 + b^2)^2}\,dx.$

17. $\displaystyle\int_0^{\infty} \frac{x \sin ax}{x^2 + b^2}\,dx.$ 18. $\displaystyle\int_0^{\infty} \frac{\cos ax}{x^4 + x^2 + 1}\,dx.$

*19. $\displaystyle\int_0^{\infty} \frac{x^{a-1}}{x^2 + 1}\,dx, \qquad 0 < a < 2.$

20. $\displaystyle\int_0^{\infty} \frac{x^{a-1}}{(z^2 + 1)^2}\,dx, \qquad 0 < a < 3.$

Prove the following results.

21. $\displaystyle\int_0^{\infty} \frac{x^{a-1}}{x^2 + x + 1}\,dx = \frac{2\pi}{\sqrt{3}} \csc \pi a \cos\left\{\frac{\pi}{6}(2a + 1)\right\}, \qquad 0 < a < 2.$

22. $\displaystyle\int_{-\infty}^{\infty} \frac{\sin^2 ax}{x^2(x^2 + b^2)}\,dx = \frac{\pi}{2b^3}(e^{-2ab} - 1 + 2ab).$

23. $\displaystyle\int_{-\infty}^{\infty} \frac{x - \sin x}{x^3(x^2 + a^2)}\,dx = \frac{\pi}{a^4}\left(\frac{1}{2}a^2 - a + 1 - e^{-a}\right).$

24. $\displaystyle\int_0^{\infty} \frac{dx}{x^{2n} + 1} = \frac{\pi}{2n} \csc \frac{\pi}{2n}$ for any positive integer n.

25. $\displaystyle\int_{-\infty + ia}^{\infty + ia} \frac{dx}{x^3 + 1} = \frac{2\pi}{3} e^{-i\pi/6}\left\{1 - u\left(a - \frac{\sqrt{3}}{2}\right)\right\},$

where u is the unit-step function.

Suggested reading

Phillips, E. G. *Functions of a Complex Variable with Applications*. 7th ed. Interscience, New York, 1951. An excellent and concise treatment.

Churchill, R. V. *Complex Variables and Applications*. McGraw-Hill, New York, 1960.

Greenberg, M. D. *Foundations of Applied Mathematics*. Prentice-Hall, Englewood Cliffs, NJ, 1978. The relevant material is covered in Chapters 12–15, with many worked examples.

CHAPTER 6

Fourier transforms

We now return to a study of linear systems and, in particular, to the steady-state response of a stable system. In Chapter 4 we saw that for a periodic input $f(t)$ the output could be obtained immediately using the Fourier-series representation of $f(t)$. For a nonperiodic $f(t)$, the development of an analogous approach leads to a consideration of Fourier transforms, which have similarities to Laplace transforms as well as to Fourier series.

6.1 Derivation from Fourier series

As shown in Section 4.7, a periodic function $f(t)$ of relatively general form can be expressed as a complex Fourier series:

$$f(t) = \sum_{n=-\infty}^{\infty} c_n e^{in\omega_0 t} \tag{6.1}$$

whose coefficients are

$$c_n = \frac{1}{\tau} \int_{-\tau/2}^{\tau/2} f(t) e^{-in\omega_0 t}\, dt, \tag{6.2}$$

where τ is a period of $f(t)$ and $\omega_0 = 2\pi/\tau$. Although it is sufficient to integrate over any interval in t of length τ, it is now convenient to take the interval to be symmetric about the origin; and to avoid confusion with a continuous variable ω that will be introduced, a subscript "0" has been added to the symbol denoting the fundamental angular frequency of $f(t)$. The coefficients c_n define the frequency spectrum of $f(t)$, and since the only frequencies involved are $n\omega_0$ for $n = 0, \pm 1, \pm 2, \ldots,$ the

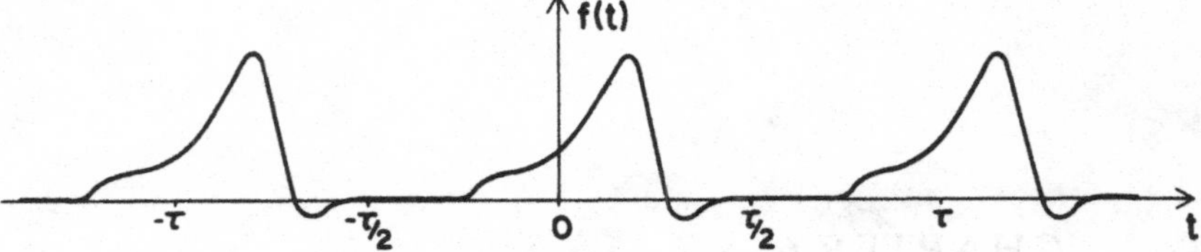

Fig. 6.1: Periodic $f(t)$ representing a pulse train.

spectrum is a discrete or line spectrum consisting of lines of "height" c_n at the frequencies $n\omega_0$. The lines are ω_0 apart and become closer as τ increases; and as τ approaches infinity the lines become so dense that we can picture the coefficients c_n defining a continuous curve representing a continuous spectrum.

With this clue, let us seek the relations analogous to (6.1) and (6.2) for a nonperiodic $f(t)$. Consider the pulse train shown in Fig. 6.1. As τ increases, a single pulse expands to fill a greater range of t and, in the limit as $\tau \to \infty$, occupies the entire span $-\infty < t < \infty$. Since the resulting function is no longer periodic, this suggests the possibility of obtaining the desired relations as a limiting form of the expressions in (6.1) and (6.2). The coefficients c_n are functions of $n\omega_0$, and if

$$F_\tau(n\omega_0) = \tau c_n = \int_{-\tau/2}^{\tau/2} f(t) e^{-in\omega_0 t}\, dt,$$

then

$$F_\tau(\omega) = \int_{-\tau/2}^{\tau/2} f(t) e^{-i\omega t}\, dt.$$

Although this is valid only for the sampled values $n\omega_0$ of ω, the samples become ever closer as τ increases, and in the limit we have

$$\lim_{\tau\to\infty} F_\tau(\omega) = F(\omega) = \int_{-\infty}^{\infty} f(t) e^{-i\omega t}\, dt,$$

where ω can now be treated as a continuous variable. Also, from Eq. (6.1),

$$f(t) = \frac{1}{2\pi} \sum_{n=-\infty}^{\infty} F_\tau(n\omega_0) e^{in\omega_0 t}\omega_0.$$

For large τ, ω_0 is a small increment $\Delta\omega$ in the continuous variable ω, and the right-hand side then represents an approximation to an infinite integral in which the integrand is sampled at the discrete values $n\omega_0$, $\Delta\omega$

apart. In the limit as $\tau \to \infty$ (i.e., $\Delta\omega \to 0$),

$$f(t) = \lim_{\Delta\omega \to 0} \frac{1}{2\pi} \sum_{n=-\infty}^{\infty} F_\tau(n\omega_0) e^{in\omega_0 t} \Delta\omega$$

$$= \frac{1}{2\pi} \int_{-\infty}^{\infty} F(\omega) e^{i\omega t} \, d\omega.$$

Thus, for a nonperiodic $f(t)$ the expression analogous to (6.1) is

$$f(t) = \frac{1}{2\pi} \int_{-\infty}^{\infty} F(\omega) e^{i\omega t} \, d\omega \tag{6.3}$$

and this is called the *Fourier-integral representation* of $f(t)$, where the analog of (6.2) is

$$F(\omega) = \int_{-\infty}^{\infty} f(t) e^{-i\omega t} \, dt. \tag{6.4}$$

By substituting (6.4) into (6.3) we obtain *Fourier's identity*:

$$f(t) = \frac{1}{2\pi} \int_{-\infty}^{\infty} \int_{-\infty}^{\infty} f(t') e^{i\omega(t-t')} \, dt' \, d\omega, \tag{6.5}$$

which also follows directly from Eq. (4.31) on taking the limit $\tau \to \infty$.

The above derivation of the identity (6.5) is purely formal and not intended as a rigorous proof, but it does have the advantage of bringing out the connection between the coefficients c_n of the Fourier series (6.1) and the weighting function $F(\omega)$ of the Fourier integral (6.3). The proof can be made rigorous (see, for example, the book by I. N. Sneddon listed at the end of this chapter), and it is also possible to derive (6.3) and (6.4) from Cauchy's integral formula (see G. Arfken).

6.2 Definition of the transforms

Transforms, like Siamese twins, come in pairs, and the basis for all integral transforms is a double integral identity similar to that shown in Eq. (6.5). Fourier's identity is the mathematical foundation for Fourier transforms, but the transforms can be defined in a variety of ways consistent with (6.5). If we choose the function $F(\omega)$ shown in Eq. (6.4), then $f(t)$ is expressed in terms of $F(\omega)$ by Eq. (6.3). Alternatively, the factor $(2\pi)^{-1}$ could be put into (6.4) instead of (6.3), or spread equally between the two by having $(2\pi)^{-1/2}$ in each equation. The latter enhances the symmetry between the two relations, but at the expense of proliferating the factors $(2\pi)^{-1/2}$ throughout all calculations involving the transforms. Another common practice is to replace the angular frequency ω by a (linear) frequency ν where $\omega = 2\pi\nu$, in which case

Fourier's identity becomes

$$f(t) = \int_{-\infty}^{\infty}\int_{-\infty}^{\infty} f(t')e^{2\pi i\nu(t-t')}\,dt'\,d\nu.$$

This transfers the factor to the exponents in both equations. If ω is replaced by $-\omega$ (or ν by $-\nu$) the identity is unaffected apart from a sign change in the exponent, and we can therefore select either sign for the exponent in the definition of $F(\omega)$ provided the exponent in the corresponding expression for $f(t)$ has the opposite sign. Clearly, some latitude exists in the definition of a Fourier-transform pair, and all of the variations discussed can be found in the literature. Before using a text or table for information about a transform, the reader is advised to check which definition is employed.

Our choice is that which is customary (but by no means universal) in linear-systems theory and is guided by the expression (6.1) for the complex Fourier-series representation of a periodic function. By definition, the *Fourier transform* of $f(t)$ is

$$\int_{-\infty}^{\infty} f(t)e^{-i\omega t}\,dt = F(\omega), \tag{6.6}$$

where ω is a continuous real variable , $-\infty < \omega < \infty$, and is written symbolically as

$$\mathscr{F}\{f(t)\} = F(\omega). \tag{6.7}$$

The corresponding transform is therefore

$$\frac{1}{2\pi}\int_{-\infty}^{\infty} F(\omega)e^{i\omega t}\,d\omega = f(t) \tag{6.8}$$

and is called the *inverse Fourier transform* of $F(\omega)$, written symbolically as

$$\mathscr{F}^{-1}\{F(\omega)\} = f(t). \tag{6.9}$$

Equations (6.6) and (6.8) together form a Fourier-transform pair consistent with the identity (6.5), and thus $\mathscr{F}^{-1}\mathscr{F}$ is an identity transformation for all functions $f(t)$ for which the transforms are valid. We shall follow the practice established in connection with the Laplace transform of using a capital (uppercase) letter to denote the transform of the function of t represented by the corresponding small (lowercase) letter. Of course, the transform in question is now the Fourier transform, and we shall deviate from this notation only when it is necessary to distinguish between the Laplace and Fourier transforms or in other special circumstances.

The Fourier transform (6.6) transforms or maps a function of t (in the time domain) into a function of ω (in the frequency domain), and the inverse Fourier transform (6.8) does the reverse. The latter can be thought of as the recovery operation whereby the function $f(t)$ is retrieved from a knowledge of its transform $F(\omega)$. Thus

$$f(t) \underset{\mathscr{F}^{-1}}{\overset{\mathscr{F}}{\rightleftarrows}} F(\omega)$$

and we shall display a Fourier-transform relationship as

$$f(t) \Leftrightarrow F(\omega). \tag{6.10}$$

Equation (6.8) is also the counterpart of (6.1) and constitutes the Fourier-integral representation of $f(t)$. $F(\omega)$ defines the *frequency spectrum* of $f(t)$. In eq. (6.8), $e^{i\omega t}$ is a single frequency oscillation of angular frequency ω, and $(2\pi)^{-1}F(\omega)$ is the weighting of each frequency ω in the integral representation of $f(t)$. In contrast to the series representation (6.1) where the coefficients c_n have the same units as $f(t)$, $(2\pi)^{-1}F(\omega)$ is actually a spectral density; that is, $(2\pi)^{-1}F(\omega)\,\Delta\omega$ is the strength of the components having frequencies in the range $(\omega - \frac{1}{2}\Delta\omega, \omega + \frac{1}{2}\Delta\omega)$.

The Fourier transform is probably the most important of all integral transforms and there are few branches of science or engineering where it does not appear. Apart from its utility as a tool for the solution of mathematical problems, the transform has a physical meaning which is based on the interpretation of $e^{i\omega t}$ as an elementary wave. The Fourier-integral representation is therefore a synthesis in terms of waves, and a Fourier-transform relationship can be exhibited using any form of wave motion, be it acoustic, optic, electromagnetic, or hydrodynamic. As an example, the Fourier transform of an illuminated aperture in optics is the Fraunhofer pattern, which is easily displayed, and in any problem involving signal analysis and/or processing, Fourier-transform methods are indispensable.

Example. Find the frequency spectrum of a rectangular pulse of length $2t_0$.

We choose the pulse to be of unit height and centered on $t = 0$ as shown in Fig. 6.2. Then

$$f(t) = u(t + t_0) - u(t - t_0) \tag{6.11}$$

and

$$F(\omega) = \mathscr{F}\{f(t)\} = \int_{-\infty}^{\infty} f(t)e^{-i\omega t}\,dt$$

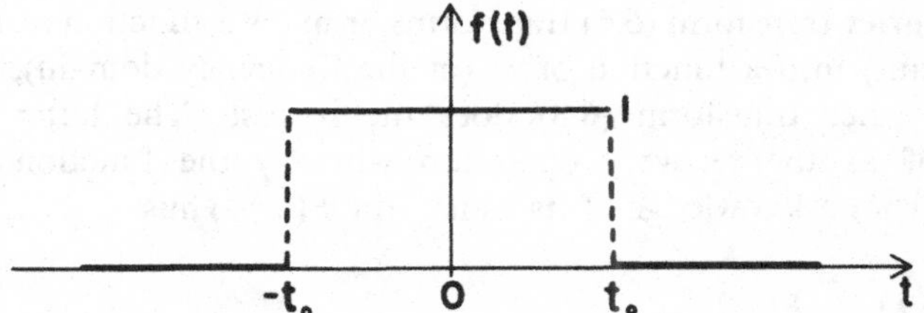

Fig. 6.2: Rectangular pulse $f(t)$.

$$= \int_{-t_0}^{t_0} e^{-i\omega t}\, dt$$

$$= \frac{1}{i\omega}\left(e^{i\omega t_0} - e^{-i\omega t_0}\right),$$

that is,

$$F(\omega) = \frac{2}{\omega}\sin \omega t_0, \tag{6.12}$$

which is a real symmetric (or even) function of ω. By l'Hospital's rule, $F(0) = 2t_0$, and a plot of $F(\omega)$ is shown in Fig. 6.3. $F(\omega)$ is zero at all nonzero integer multiples of π/t_0, and although $F(\omega)$ extends out to $\pm\infty$, the dominant part of the spectrum lies in the range $0 \leq |\omega| \lesssim \pi/t_0$. The range increases as t_0 decreases, showing that the frequency bandwidth necessary to produce a rectangular pulse is inversely proportional to the pulse width. If $t_0 = \frac{1}{2}$ the pulse has unit area and, depending on the discipline, is referred to as a *gate*, *aperture*, or *slit function*. A notation that is sometimes adopted is

$$\operatorname{rect}(t) = u\left(t + \tfrac{1}{2}\right) - u\left(t - \tfrac{1}{2}\right), \tag{6.13}$$

in which case, from (6.12),

$$\operatorname{rect}(t) \Leftrightarrow \frac{\sin \omega/2}{\omega/2}. \tag{6.14}$$

6.3 Conditions for existence

Up to this point we have not considered the validity of the transform operations defined in Eqs. (6.6) and (6.8), and have merely assumed that they are meaningful for the functions of interest. To an optical engineer working with fringe patterns or to a circuit designer whose acquaintance with Fourier transforms is through practical experience, physical possibility is a sufficient condition for the existence of a transform, and as long

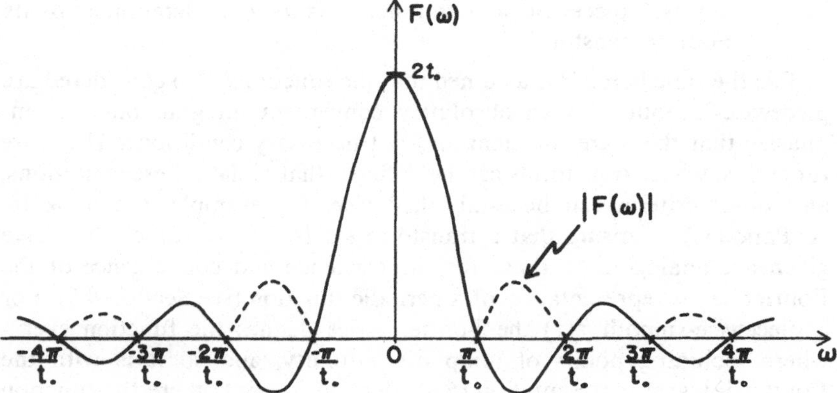

Fig. 6.3: Frequency spectrum of the rectangular pulse $f(t)$.

as the function accurately describes a physical quantity, nothing more need be said. Nevertheless, it is often convenient to replace the function by one whose mathematical form is simpler, and such functions as $\sin t$ and $u(t)$ do not have Fourier transforms in the strictest sense. Practically as well as mathematically it is therefore important to have conditions that guarantee the existence of the transform.

The Fourier transform is an infinite-integral operator applied to the function $f(t)$, and as we remarked in the case of the Laplace transform, not all functions can be integrated over even a finite range, and not all infinite integrals converge. The following conditions address both of these matters, and proofs can be found in the book by W. Kaplan.

(i) If $f(t)e^{-i\omega t}$ is integrable for $|t| < T < \infty$ (it is sufficient if $f(t)$ is *piecewise-continuous* in this range), and

(ii) if $\int_{-\infty}^{\infty} f(t)\, dt$ is *absolutely convergent*, that is,

$$\int_{-\infty}^{\infty} |f(t)|\, dt < \infty,$$

implying $f(t) \to 0$ as $t \to \pm\infty$, the transform integral (6.6) is meaningful, and $F(\omega)$ is defined for all ω, $-\infty < \omega < \infty$.

(iii) If, in addition, $f(t)$ is *piecewise-smooth* for $|t| < T < \infty$ (so that its right- and left-hand derivatives exist at every point), $F(\omega)$ is a continuous function of ω, and the inverse transform integral (6.8) converges for all t, implying $F(\omega) \to 0$ as $\omega \to \pm\infty$. The value of the integral is $F(t)$ wherever $f(t)$ is continuous, and the average of the right- and left-hand limits at any point of jump discontinuity in $f(t)$. Except possibly at such points, a piecewise-smooth function is *uniquely* determined by its Fourier transform.

For the time being it is assumed that the functions $f(t)$ considered are piecewise-smooth with an absolutely convergent integral, but we emphasize that these are sufficient and not necessary conditions. There are functions, whose transforms can be defined, that violate these conditions, and other criteria can be established (see, for example, the book by A. Papoulis) to ensure that a transform exists. The conditions we have given are analogous to those for the existence and convergence of the Fourier-series representation of a periodic function (see Section 4.5). For a piecewise-smooth $f(t)$ the Fourier series defines the function everywhere including points of jump discontinuity, and so it is with the Fourier-integral representation (6.8). Even at a point where the function $f(t)$ used to generate the transform $F(\omega)$ is undefined, the inverse transform converges to the average of the right- and left-hand limits, thereby specifying $f(t)$.

6.4 Duality, symmetry, and other properties

There is one result that should be pointed out immediately. From the definition (6.6)

$$\begin{aligned}\mathscr{F}\{f(-t)\} &= \int_{-\infty}^{\infty} f(-t)e^{-i\omega t}\,dt \\ &= \int_{-\infty}^{\infty} f(t')e^{i\omega t'}\,dt' \qquad t' = -t \\ &= F(-\omega)\end{aligned}$$

and thus (6.10) implies

$$f(-t) \Leftrightarrow F(-\omega). \tag{6.15}$$

If the definitions of the direct and inverse transforms are compared, it is observed that apart from the factor 2π the integral operators differ only in the signs of the exponents. This similarity between the two operations is called duality, and it has several consequences. From Eq. (6.8) the inverse transform can be written as

$$f(t) = \int_{-\infty}^{\infty} \frac{1}{2\pi}F(\omega)e^{-i\omega(-t)}\,d\omega,$$

showing that

$$\frac{1}{2\pi}F(t) \Leftrightarrow f(-\omega) \tag{6.16}$$

and similarly

$$\frac{1}{2\pi}F(-t) \Leftrightarrow f(\omega). \tag{6.17}$$

If follows that from any transform pair either one or three additional pairs can be established, and in particular, from (6.14)

$$\frac{\sin t/2}{t/2} \Leftrightarrow 2\pi\,\mathrm{rect}(\omega). \tag{6.18}$$

Thus, $(\sin t/2)/(t/2)$ is a band-limited function whose frequency spectrum is zero for $|\omega| > \frac{1}{2}$, but since its integral does not converge absolutely, it is also an example of a function that violates condition (ii) of Section 6.3 yet still has a Fourier transform.

Another consequence of duality is that the inverse transform operation can be expressed as a direct transform. If a bar denotes the complex conjugate, Eq. (6.6) implies that

$$\bar{f}(t) = \frac{1}{2\pi}\int_{-\infty}^{\infty} \bar{F}(\omega)e^{-i\omega t}\,d\omega$$

and hence

$$\mathscr{F}^{-1}\{F(\omega)\} = \overline{\mathscr{F}\left\{\frac{1}{2\pi}\bar{F}(\omega)\right\}}\Bigg|_{\omega\mapsto t}, \tag{6.19}$$

which constitutes an alternative inversion formula. This is important for numerical purposes and is used in most fast Fourier-transform (FFT) computer programs.

As evident from Eq. (6.6), the Fourier transform is, in general, a complex function of ω, and this is true even if $f(t)$ is real, but any symmetry that $f(t)$ possesses may serve to restrict $F(\omega)$. By expanding the exponential in (6.6) we have

$$F(\omega) = \int_{-\infty}^{\infty} f(t)\cos\omega t\,dt - i\int_{-\infty}^{\infty} f(t)\sin\omega t\,dt. \tag{6.20}$$

If $f(t)$ is an *even* function [i.e., $f(-t) = f(t)$], the second integral in (6.20) vanishes and

$$F(\omega) = 2\int_0^{\infty} f(t)\cos\omega t\,dt. \tag{6.21}$$

Thus, $F(\omega)$ is an *even* function that is real if $f(t)$ is real, and from Eq. (6.8),

$$f(t) = \frac{1}{\pi}\int_0^{\infty} F(\omega)\cos\omega t\,d\omega. \tag{6.22}$$

Table 6-1. *Symmetry properties of the transforms*

Time domain $f(t)$	Frequency domain $F(\omega)$
Real	Real part even Imaginary part odd
Real and even	Real and even
Real and odd	Imaginary and odd
Imaginary	Real part odd Imaginary part even
Imaginary and even	Imaginary and even
Imaginary and odd	Real and odd
Complex and even	Complex and even
Complex and odd	Complex and odd
Real even, imaginary odd	Real
Real odd, imaginary even	Imaginary

This representation of an even function is called a *Fourier cosine integral*, and (6.21) and (6.22) constitute a pair of transform relations called *Fourier cosine transforms*. Similarly, if $f(t)$ is an *odd* function [i.e., $f(-t) = -f(t)$], the first integral in (6.20) vanishes and

$$F(\omega) = -2i\int_0^\infty f(t)\sin\omega t\,dt. \tag{6.23}$$

$F(\omega)$ is then an *odd* function that is purely imaginary if $f(t)$ is real, and from Eq. (6.8),

$$f(t) = \frac{i}{\pi}\int_0^\infty F(\omega)\sin\omega t\,d\omega. \tag{6.24}$$

This is a *Fourier-sine-integral* representation of an odd function of t, and the transform relations (6.23) and (6.24) are called *Fourier sine transforms*. These symmetry properties are listed in Table 6-1.

Though we shall have no cause to use the cosine and sine transforms per se, we remark that if $f(t)$ is defined only for $0 \leq t < \infty$, Eqs. (6.22) and (6.24) provide representations of the even and odd continuations, respectively, over the entire range $-\infty < t < \infty$.

From the definition (6.6) of the Fourier transform it follows immediately that

$$F(0) = \int_{-\infty}^{\infty} f(t)\,dt \tag{6.25}$$

and similarly

$$f(0) = \frac{1}{2\pi}\int_{-\infty}^{\infty} F(\omega)\,d\omega. \tag{6.26}$$

Thus, $F(0)$ is equal to the integral of the function in the time domain, a result that is analogous to the interpretation of the coefficient c_0 in the Fourier-series expansion of a periodic function, and $f(0)$ is likewise equal to the integral of the spectral density in the frequency domain.

Parseval's formula

In the time domain the energy in a signal $f(t)$ is defined as

$$E = \int_{-\infty}^{\infty} |f(t)|^2 \, dt \tag{6.27}$$

and if, for example, $f(t)$ is the voltage drop across a 1 ohm resistance, E is the total energy delivered by the source. Using Eq. (6.8),

$$\begin{aligned} E &= \int_{-\infty}^{\infty} \bar{f}(t) f(t) \, dt \\ &= \frac{1}{2\pi} \int_{-\infty}^{\infty} \bar{f}(t) \left(\int_{-\infty}^{\infty} F(\omega) e^{i\omega t} \, d\omega \right) dt \\ &= \frac{1}{2\pi} \int_{-\infty}^{\infty} F(\omega) \left(\int_{-\infty}^{\infty} \bar{f}(t) e^{i\omega t} \, dt \right) d\omega \end{aligned}$$

on reversing the order of integration. But from Eq. (6.6)

$$\int_{-\infty}^{\infty} \bar{f}(t) e^{i\omega t} \, dt = \bar{F}(\omega),$$

and hence

$$E = \int_{-\infty}^{\infty} |f(t)|^2 \, dt = \frac{1}{2\pi} \int_{-\infty}^{\infty} |F(\omega)|^2 \, d\omega. \tag{6.28}$$

This is often called *Parseval's formula* (or theorem), after the French mathematician Marc Antoine Parseval des Chenes (1755–1836) who derived the analogous result for Fourier series. Another name is *Plancherel's theorem* in recognition of the Swiss mathematician Michel Plancherel (1885–), whose 1910 paper established conditions under which the theorem is true, but it appears that the first publication of Eq. (6.28) was by the English physicist Lord Rayleigh (1842–1919) in an 1897 paper. Regardless of its parentage, the equation is obvious. Since $1/2\pi|F(\omega)|^2$ is the energy spectral density, integration over all frequencies gives the total energy contained in the signal, which must be the same in the time and frequency domains.

6.5 Evaluation of the transforms

Analytical evaluation

The direct and inverse Fourier transforms are mathematically similar operations, and since the methods for evaluating them are the same, it is sufficient to consider the direct transform only.

For most functions $f(t)$ the transform integral is difficult to evaluate analytically and, prior to the development of the FFT algorithm, was also difficult to treat numerically, but there are still functions of practical interest whose form is simple enough for us to evaluate the integrals. Needless to say, they are the ones considered in the examples and exercises, and for a subset of these, path closure is possible in the complex t plane, leading to an evaluation in terms of residues. Clearly, $f(t)$ must be meromorphic throughout the entire plane, and since an analytic function is continuous with all derivatives continuous, it cannot involve $|t|$ or the unit-step function $u(t)$. When coupled with the requirement that $f(t) \to 0$ as $|t| \to \infty$, the subset consists of functions that are proper rational functions of t, or the product of proper rational functions and exponents of the form e^{iat} with a real. For all other $f(t)$ the integral must be evaluated directly. We shall illustrate both methods.

Example 1. Evaluate $\mathscr{F}\{f(t)\}$ with $f(t) = e^{2it}/(t^2+1)$.

The Fourier transform is

$$F(\omega) = \int_{-\infty}^{\infty} \frac{e^{2it}}{t^2+1} e^{-i\omega t}\, dt = \int_{-\infty}^{\infty} g(z)\, dz,$$

where

$$g(z) = \frac{e^{-i(\omega-2)z}}{(z-i)(z+i)}.$$

Since

$$e^{-i(\omega-2)z} = e^{-i(\omega-2)x} e^{(\omega-2)y},$$

the requirement that $(\omega-2)y$ be negative in the half-plane of closure shows that

$$\text{if} \quad \omega - 2 < 0, \quad \text{then} \quad y > 0 \text{ (UHP)},$$

whereas

$$\text{if} \quad \omega - 2 > 0, \quad \text{then} \quad y < 0 \text{ (LHP)}.$$

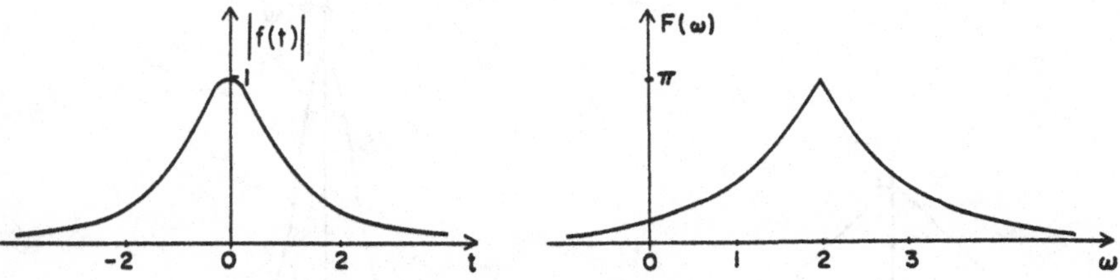

Fig. 6.4: The magnitude of $f(t)$ and its frequency spectrum $F(\omega)$.

The conditions for path closure (see Section 5.7) are therefore satisfied in the half-planes indicated.

(i) If $\omega - 2 < 0$,

$$F(\omega) = 2\pi i \operatorname{Res}\{f(z), i\}$$

$$= 2\pi i \frac{e^{\omega-2}}{2i} = \pi e^{\omega-2};$$

(ii) if $\omega - 2 > 0$,

$$F(\omega) = -2\pi i \operatorname{Res}\{g(z), -i\}$$

$$= -2\pi i \frac{e^{-(\omega-2)}}{-2i} = \pi e^{-(\omega-2)}.$$

Hence

$$F(\omega) = \pi e^{-|\omega-2|}$$

for all ω, and

$$\frac{e^{2it}}{t^2+1} \Leftrightarrow \pi e^{-|\omega-2|}. \tag{6.29}$$

Although the case in which $\omega = 2$ was excluded from the above evaluation, separate consideration is unnecessary, since $F(\omega)$ must be continuous for all ω.

The function $f(t)$ and its frequency spectrum are shown in Fig. 6.4, and we observe that both are infinite in extent. We note that because of the discontinuity in the derivative $F'(\omega)$ at $\omega = 2$ produced by the modulus, it is not possible to use the residue method to evaluate $\mathscr{F}^{-1}\{F(\omega)\}$.

Example 2. Find the frequency spectrum of the triangular pulse $f(t) = \operatorname{tri}(t)$ where

$$\operatorname{tri}(t) = (1 - |t|)\{u(t+1) - u(t-1)\}. \tag{6.30}$$

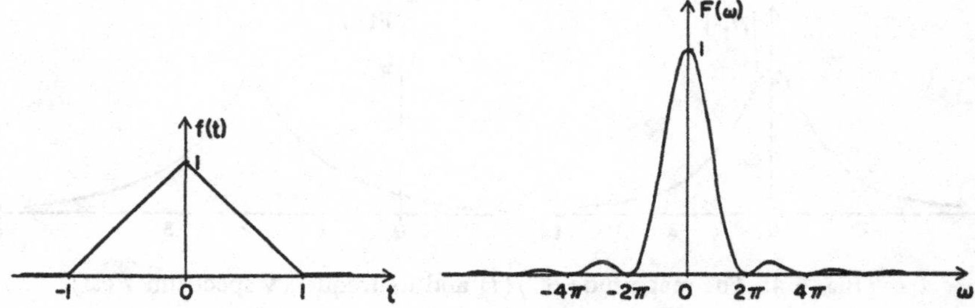

Fig. 6.5: The triangular pulse function $f(t) = \text{tri}(t)$ and its frequency spectrum.

As indicated by the presence of the step function, the Fourier transform must be evaluated directly. We have

$$F(\omega) = \int_{-\infty}^{\infty} (1 - |t|)\{u(t+1) - u(t-1)\} e^{-i\omega t}\, dt$$

$$= \int_{-1}^{0} (1+t) e^{-i\omega t}\, dt + \int_{0}^{1} (1-t) e^{-i\omega t}\, dt$$

and integration by parts gives

$$F(\omega) = \left[\frac{1}{-i\omega}\left(1 + t + \frac{1}{i\omega}\right)e^{-i\omega t}\right]_{-1}^{0} + \left[\frac{1}{-i\omega}\left(1 - t - \frac{1}{i\omega}\right)e^{-i\omega t}\right]_{0}^{1}$$

$$= \frac{2}{\omega^2}(1 - \cos\omega) = \frac{4}{\omega^2}\sin^2\frac{\omega}{2}.$$

Thus

$$\text{tri}(t) \Leftrightarrow \left(\frac{\sin\omega/2}{\omega/2}\right)^2. \tag{6.31}$$

The triangular pulse and its frequency spectrum are shown in Fig. 6.5. Like the rectangular pulse, the signal has finite duration and though its spectrum extends to infinity, the oscillations decrease much more rapidly than those for $\text{rect}(t)$ (see Fig. 6.3). Since $F(0) = 1$, Eq. (6.25) is clearly satisfied, and since $f(0) = 1$, Eq. (6.26) implies

$$\int_{-\infty}^{\infty} \frac{\sin^2\omega}{\omega^2}\, d\omega = \pi.$$

Finally, we note that $\mathscr{F}^{-1}\{F(\omega)\}$ can be evaluated using the residue method (see Exercise 5 following).

Example 3. Evaluate $\mathscr{F}\{e^{-t}u(t)\}$.

It is obvious that the transform must be evaluated directly, and the task is trivial. We have

$$\mathscr{F}\{e^{-t}u(t)\} = \int_0^\infty e^{-t}e^{-i\omega t}\,dt$$
$$= \left[\frac{e^{-(i\omega+1)t}}{-(i\omega+1)}\right]_0^\infty = \frac{1}{i\omega+1}$$

and thus

$$e^{-t}u(t) \Leftrightarrow \frac{1}{i\omega+1}. \tag{6.32}$$

Exercises

It is intended that the solutions be obtained by evaluating the appropriate transform integrals, without reference to any specific result or property derived heretofore.

Find the inverse Fourier transforms of the following $F(\omega)$.

*1. $e^{-|\omega|}$.

2. $u(\omega) - u(\omega - \omega_0) + u(\omega - 2\omega_0) - u(\omega - 3\omega_0)$.

3. $\dfrac{1}{(\omega - i)^2 + 4}$.

4. $\dfrac{\sin \omega t_0}{(\omega - i)^2}$.

5. $\left(\dfrac{\sin \omega/2}{\omega/2}\right)^2$ (see Example 3 on p. 199).

Find the frequency spectra of the following $f(t)$.

*6. $\left(1 - \dfrac{t}{t_0}\right)\{u(t) - u(t - t_0)\}$.

7. $e^{-t}(\sin \omega_0 t)u(t)$.

8. $\dfrac{1}{(t^2 + a^2)^2}$, $\quad \operatorname{Re} a > 0$.

9. $\left(\sin\dfrac{2\pi t}{t_0}\right)\{u(t) - u(t - t_0)\}$.

10. $\cos t \cos 9t\{u(t + \pi/2) - u(t - \pi/2)\}$.

Find the Fourier-integral representation of the following $f(t)$.

*11. $(1 + t)\{u(t + 1) - u(t)\} - (1 - t)\{u(t) - u(t - 1)\}$.

12. $(\sin t)\{u(t) - u(t - \pi)\}$.

13. $te^{-|t|}$.

14. $te^{-t}u(t)$.

15. $\dfrac{t \cos t}{t^2 + 1}$.

16. For the waveforms shown in Fig. 6.6 express $F_1(\omega)$ in terms of $F_2(\omega)$.

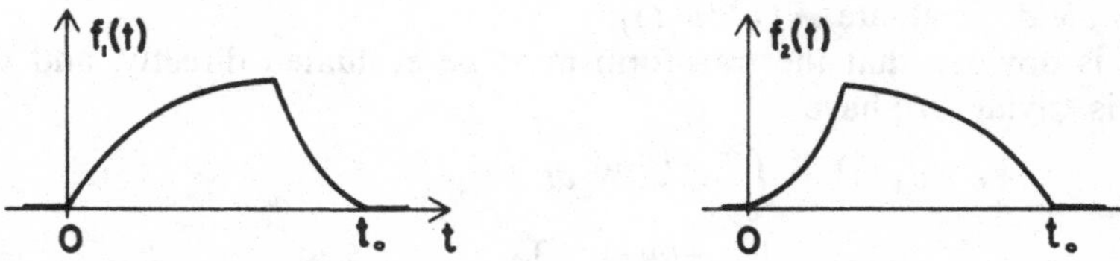

Fig. 6.6: Functions $f_1(t)$ and $f_2(t)$ in Exercise 16.

Numerical evaluation

The simple examples discussed so far have involved functions whose Fourier transforms can be obtained analytically. Although combinations of these functions may be used to approximate waveforms obtained experimentally, a less time-consuming method in this case is to sample the data and then compute an approximation of the Fourier transform with a high-speed digital computer. To this end a numerical method of approximating the Fourier transform is now examined.

The Fourier transform and its inverse are

$$F(\omega) = \int_{-\infty}^{\infty} f(t) e^{-i\omega t}\, dt,$$

$$f(t) = \frac{1}{2\pi} \int_{-\infty}^{\infty} F(\omega) e^{-i\omega t}\, d\omega,$$

as given in Section 6.2. Both integrals may be approximated using a piecewise-constant curve fit to obtain

$$\begin{aligned} F(\omega) &\cong F_1(\omega) = \sum_{n=-\infty}^{\infty} f(n\,\Delta t) e^{-in(\Delta t)\omega} \Delta t, \\ f(t) &\cong f_1(t) = \frac{1}{2\pi} \sum_{m=-\infty}^{\infty} F(m\,\Delta\omega) e^{im(\Delta\omega)t} \Delta\omega. \end{aligned} \tag{6.33}$$

The accuracy of the approximation depends on the smoothness of the functions $f(t)$ and $F(\omega)$ and the extent to which the sampling of these functions is sufficiently fine. For example, if $f(t)$ has a 20-Hz component and we sample $f(t)$ at intervals $\Delta t = 0.2$ seconds corresponding to 5 Hz, the numerical integration of the 20-Hz component would not be expected to be accurate.

An interesting consequence of the approximation given by (6.33) is that $F_1(\omega)$ and $f_1(t)$ are both periodic. As shown in Section 4.7, if

$$f(t) = \sum_{n=-\infty}^{\infty} c_n e^{in(\Delta\omega)t}$$

then $f(t)$ is periodic with period $\tau = 2\pi/\Delta\omega$. Thus, $f_1(t)$ has period $T = 2\pi/\Delta\omega$ and $F_1(\omega)$ has period $W = 2\pi/\Delta t$. The periodicity is, of course, only due to the numerical approximations, and since we are discussing the Fourier transform (not the Fourier series), the functions $F(\omega)$ and $f(t)$ must be assumed nonperiodic. The idea of approximating a nonperiodic function with a periodic function is rather unsettling, and the inconsistency is resolved by using $F_1(\omega)$ and $f_1(t)$ to approximate $F(\omega)$ and $f(t)$ only within the first period. Outside of this first period, $F(\omega)$ and $f(t)$ are approximated as zero. Since a period that is too short may thus have an adverse effect on the accuracy of the transformation, it is important that $f(t) \cong 0$ for $t > T\,(= 2\pi/\Delta\omega)$ and $t < 0$, and $F(\omega) \cong 0$ for $2|\omega| > W\,(= 2\pi/\Delta t)$. Choosing Δt and $\Delta\omega$ too large will result in periods that are too small and will produce errors known as *aliasing errors*. On the other hand, choosing $\Delta\omega$ and Δt too small will result in periods that are larger than necessary and is computationally inefficient.

Since $F(\omega)$ and $f(t)$ are assumed zero for $|\omega| > W/2$ and $t > T$ and $t < 0$, (6.33) may be reduced to

$$
\begin{aligned}
F(\omega) \cong F_2(\omega) &= \sum_{n=0}^{N-1} f(n\,\Delta t)e^{-in(\Delta t)\omega}\,\Delta t, \\
f(t) \cong f_2(t) &= \frac{1}{2\pi}\sum_{m=-(M-1)/2}^{(M-1)/2} F(m\,\Delta\omega)e^{im(\Delta\omega)t}\,\Delta\omega,
\end{aligned}
\tag{6.34}
$$

where $N\Delta t = T$, $M\Delta\omega = W$, and $M - 1$ is, for the moment, assumed even. The $n = N$ and $m = (M+1)/2$ terms are omitted from the summation since $f_1(t)$ and $F_1(\omega)$ will be approximated as periodic $[f_1(0) = f_1(N)]$ so that the inclusion of these terms would imply summing the same point twice. To make these periodic assumptions explicit, the original functions $F(\omega)$ and $f(t)$ in (6.34) are replaced by their periodic approximations, $F_2(\omega)$ and $f_2(t)$, yielding

$$
\begin{aligned}
F_2(\omega) &= \Delta t \sum_{n=0}^{N-1} f_2(n\,\Delta t)e^{-in(\Delta t)\omega}, \\
f_2(t) &= \frac{\Delta\omega}{2\pi}\sum_{m=0}^{M-1} F_2(m\,\Delta\omega)e^{im(\Delta\omega)t},
\end{aligned}
\tag{6.35}
$$

where the limits of summation in the latter equation were modified from those in (6.34) by employing the periodicity of $F_2(\omega)$. In practice, this means that when one digitizes a signal over, let us say, a time interval T, the transform obtained will correspond to a periodic version of the sampled signal; that is,

$$f_2(t) = \sum_{n=-\infty}^{\infty} f(t+nT).$$

Depending on the use of the transformed signal, it may be desirable to pad the original signal with zeros. For example, if $f(t)$ is measured at $t = 0, 1, 2, \ldots, 20$, then T might be chosen as 40 instead of 20 with $f_2(t) = f(t)$ for $t = 0, 1, 2, \ldots, 20$ and $f_2(t) = 0$ for $t = 21, 22, \ldots, 40$.

Although (6.35) constitutes a valid discrete Fourier transform and inverse, it is customary to rewrite the equations using the following substitutions to facilitate computer implementation:

$$\omega = m\,\Delta\omega \quad \text{and} \quad t = n\,\Delta t$$
$$F_2(\omega) = F_2(m\,\Delta\omega) = \hat{F}(m)\,\Delta t$$
$$f_2(t) = f_2(n\,\Delta t) = \hat{f}(n)/\Delta t.$$

Further, M is set equal to N. Since $T = N\Delta t = 2\pi/\Delta\omega$ and $W = N\Delta\omega = 2\pi/\Delta t$, it follows that $\Delta t\,\Delta\omega = 2\pi/N$, which yields

$$\begin{aligned}\hat{F}(m) &= \sum_{n=0}^{N-1} \hat{f}(n) e^{-imn2\pi/N},\\ \hat{f}(n) &= \frac{1}{N} \sum_{m=0}^{N-1} \hat{F}(m) e^{imn2\pi/N}.\end{aligned} \tag{6.36}$$

The preceding two equations constitute the discrete Fourier-transform pair. One of the main reasons the discrete Fourier transform has proved so popular is the existence of highly efficient computer programs for performing the transform. As opposed to a straightforward implementation of (6.36), which would involve calculating the individual points of the transform signal in a sequential manner, the fast Fourier transform computes the elements of the transform signal in parallel, with an associated dramatic decrease in the required computation time. There are many different fast Fourier-transform (or FFT) routines available, but all are based on the discrete Fourier-transform pair (6.36).

Example. To illustrate the discrete Fourier transform, we will compute the transform of $f(t) = u(t) - u(t - t_0)$. First, calculating the continuous Fourier transform,

$$\begin{aligned}F(\omega) &= \int_{-\infty}^{\infty} f(t) e^{-i\omega t}\,dt = \int_0^{t_0} e^{-i\omega t}\,dt = \left[\frac{1}{-i\omega} e^{-i\omega t}\right]_0^{t_0}\\ &= \frac{i}{\omega}(e^{-i\omega t_0} - 1) = t_0 e^{-i\omega t_0/2} \frac{\sin(\omega t_0/2)}{\omega t_0/2},\end{aligned} \tag{6.37}$$

in agreement with the time-shift and scaling properties applied to (6.12).

For the discrete Fourier transform, given by the first equation of (6.36), we will evaluate the transform with all of the parameters of the transformation kept arbitrary:

$$\hat{F}(m) = \sum_{n=0}^{N-1} \hat{f}(n) e^{-imn2\pi/N}$$

$$= \sum_{n=0}^{L-1} e^{-imn2\pi/N},$$

where $L/N = t_0/T$ so that $\hat{f}(n) = 1$ for $0 \leq n \leq L-1$, $\hat{f}(n) = 0$ for $L-1 < n \leq N-1$, and L is assumed less than or equal to N. Noting that

$$(1 - e^{-im2\pi/N}) \sum_{n=0}^{L-1} e^{-imn2\pi/N} = \sum_{n=0}^{L-1} e^{-imn2\pi/N} - \sum_{n=1}^{L} e^{-imn2\pi/N}$$

$$= 1 - e^{-imn2\pi L/N},$$

then

$$\hat{F}(m) = \sum_{n=0}^{L-1} e^{-imn2\pi/N} = \frac{1 - e^{-im2\pi L/N}}{1 - e^{-im2\pi/N}} = \frac{e^{-im\pi L/N}}{e^{-im\pi/N}} \frac{\sin(m\pi L/N)}{\sin(m\pi/N)}. \tag{6.38}$$

Using $\Delta t = t_0/L$ and $\omega = m2\pi/T$, this becomes

$$F(m2\pi/T) \cong \hat{F}(m)\,\Delta t$$

$$= e^{im\pi/N} \frac{m\pi/N}{\sin(m\pi/N)} \left\{ t_0 e^{-im\pi L/N} \frac{\sin(m\pi L/N)}{m\pi L/N} \right\}, \tag{6.39}$$

and with ω evaluated at $m2\pi/T$, (6.37) becomes

$$F(m2\pi/T) = t_0 e^{-im\pi L/N} \frac{\sin(m\pi L/N)}{m\pi L/N}. \tag{6.40}$$

Equation (6.39), which was obtained using the discrete Fourier transform, may thus be used as an approximation to (6.40), which was obtained using the continuous Fourier transform. The approximation is accurate only if $m\pi/N \ll 1$, or equivalently $\omega \ll 2L/t_0$.

Exercises

1. The error associated with approximating (6.40) by (6.39) may be measured by defining the percent error as $|(\hat{F}\Delta t - F)/F| \times 100$.

Show that for small values of $m\pi/N$ this may be approximated as $100 \times m\pi/N$.

2. Prove that (6.39) is periodic with period $m = N$.

3. If (6.39) is to be used to approximate (6.40) for $\omega = 0, 0.1, 0.2, 0.3 \ldots, 10$ radians per second with no more than 0.1 percent error, find N, L, and T assuming $t_0 = 10\pi$ seconds. Which values of ω are associated with the maximum and minimum error?

4. In the preceding example, if L is set equal to N, then $\hat{f}(n) = 1$ for all n. Equation (6.38) simplifies for this case. What is $\hat{F}(m)$ for $N = 10$? 100? 1000? Do you see a pattern? This phenomenon can be explained in terms of delta functions, which will be discussed in Section 6.9.

5. Prove that the discrete Fourier transform and the inverse discrete Fourier transform are inverses. Do this by substitution using (6.36).

6. We have derived the discrete Fourier transform for functions of time that are zero for $t > T$ and $t < 0$. Rederive the discrete Fourier transform for functions of time that are zero for $|t| > T/2$.

6.6 Properties of the Fourier transform

The results of the preceding examples and exercises can be used to create a table of Fourier transforms listing the functions most often encountered, but no matter how complete the table, there will always be functions that are not included. In the case of the Laplace transform we were able to get by with only a short list of memorable transforms by using certain general properties that were established, and the Fourier transform has similar properties. More important, because the Fourier transform defines the frequency spectrum of a function and is a physically meaningful quantity, general properties can show the effects that various modifications to a signal have on its spectrum. Some of the required properties were described in Section 6.4, and we present here only those that are analogous to the Laplace-transform properties discussed in Sections 2.5 and 2.6 and which are most useful in practice. Unless otherwise stated it is assumed that $f(t)$ is piecewise-smooth with an absolutely convergent integral in $-\infty < t < \infty$.

Linearity

$$\mathscr{F}\{c_1 f_1(t) + c_2 f_2(t)\} = c_1 F_1(\omega) + c_2 F_2(\omega), \tag{6.41}$$

where c_1 and c_2 are constants.

Proof: This follows immediately from the linearity of the integral operator, and the extension to any finite number of terms is obvious. □

Scaling (or similarity)

$$\mathscr{F}\{f(at)\} = \frac{1}{|a|}F\left(\frac{\omega}{a}\right) \tag{6.42}$$

for any real a.

Proof: If $a > 0$,

$$\begin{aligned}\mathscr{F}\{f(at)\} &= \int_{-\infty}^{\infty} f(at)e^{-i\omega t}\,dt \\ &= \frac{1}{a}\int_{-\infty}^{\infty} f(t')e^{-i(\omega/a)t'}\,dt' \qquad t' = at \\ &= \frac{1}{a}F\left(\frac{\omega}{a}\right).\end{aligned}$$

If $a < 0$, the sign of the end result is changed because the limits of integration are interchanged. The particular case $a = -1$ corresponds to a time reversal and shows that the spectrum is also reversed [see Eq. (6.15)]. □

Frequency shift

$$\mathscr{F}\{e^{iat}f(t)\} = F(\omega - a) \tag{6.43}$$

for any real a.

Proof:

$$\begin{aligned}\mathscr{F}\{e^{iat}f(t)\} &= \int_{-\infty}^{\infty} e^{iat}f(t)e^{-i\omega t}\,dt \\ &= \int_{-\infty}^{\infty} f(t)e^{-i(\omega - a)t}\,dt \\ &= F(\omega - a). \quad □\end{aligned}$$

Example. Find the frequency spectrum of

$$f(t) = \cos\omega_0 t\{u(t + t_0) - u(t - t_0)\}.$$

From Eq. (6.12)

$$\mathscr{F}\{u(t + t_0) - u(t - t_0)\} = \frac{2}{\omega}\sin\omega t_0$$

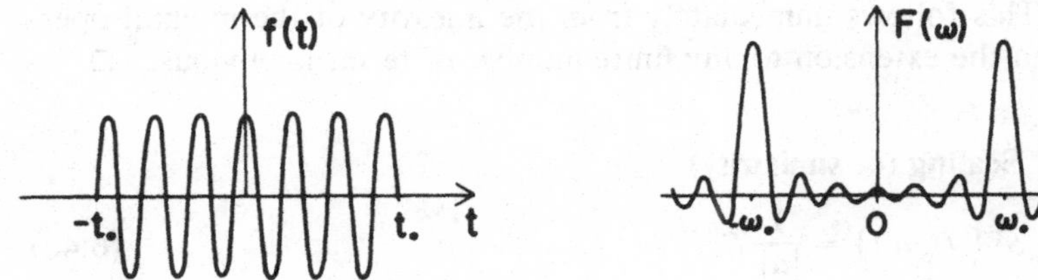

Fig. 6.7: The function $f(t)$ and its spectrum for $2\omega_0 t_0 = 13\pi$.

and hence

$$\begin{aligned}
&\mathscr{F}\{\cos\omega_0 t[u(t+t_0) - u(t-t_0)]\} \\
&\quad = \frac{1}{2}\mathscr{F}\{(e^{i\omega_0 t} + e^{-i\omega_0 t})[u(t+t_0) - u(t-t_0)]\} \\
&\quad = \frac{\sin\{(\omega-\omega_0)t_0\}}{\omega-\omega_0} + \frac{\sin\{(\omega+\omega_0)t_0\}}{\omega+\omega_0}.
\end{aligned}$$

The function and its spectrum are shown in Fig. 6.7 for the case $2\omega_0 t_0 = 13\pi$, and we observe that multiplication by $\cos\omega_0 t$ has shifted the spectrum of the original pulse so that one half is centered on $\omega = \omega_0$ and the other half on $\omega = -\omega_0$. This is the basis for a mathematical understanding of amplitude modulation. If $f(t) = g(t)\cos\omega_0 t$, $g(t)$ can be regarded as a modulation applied to the (co)sinusoidal carrier $\cos\omega_0 t$, and the spectrum of $f(t)$ is then

$$F(\omega) = \tfrac{1}{2}\{G(\omega-\omega_0) + G(\omega+\omega_0)\}.$$

Time shift

$$\mathscr{F}\{f(t-a)\} = e^{-i\omega a}F(\omega) \tag{6.44}$$

for any real a.

Proof:

$$\begin{aligned}
\mathscr{F}\{f(t-a)\} &= \int_{-\infty}^{\infty} f(t-a)e^{-i\omega t}\,dt \\
&= \int_{-\infty}^{\infty} f(t')e^{-i\omega(t'+a)}\,dt' \qquad t' = t-a \\
&= e^{-i\omega a}F(\omega).
\end{aligned}$$

Thus, a time shift has no effect on the magnitude of the spectrum but changes its phase by ωa. It is the dual of a frequency shift. □

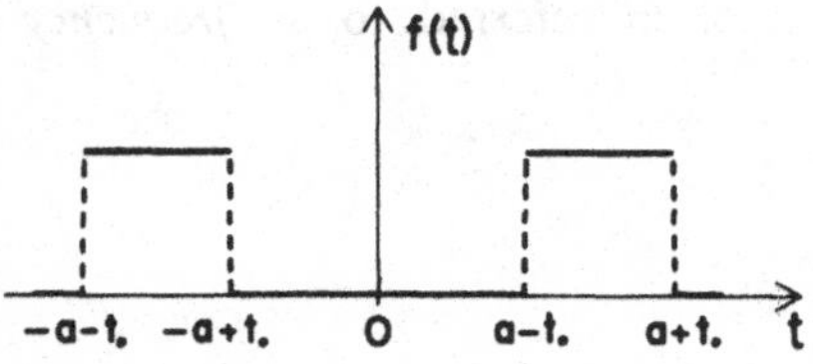

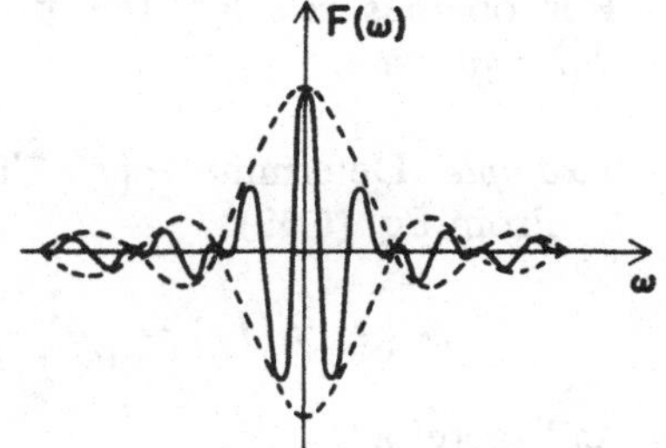

Fig. 6.8: The function $f(t)$ and its spectrum for $a = 3t_0$.

Example. Find the frequency spectrum of

$$f(t) = \tfrac{1}{2}\{u(t+a+t_0) - u(t+a-t_0) + u(t-a+t_0) - u(t-a-t_0)\}$$

for $a \gg t_0$.

From Eq. (6.12)

$$F(\omega) = \frac{1}{2}(e^{i\omega a} + e^{-i\omega a})\frac{2}{\omega}\sin \omega t_0$$

$$= \frac{2}{\omega}\sin \omega t_0 \cos \omega a.$$

Since $a \gg t_0$, the spectrum is a rapid oscillation within the envelope $\pm 2(\sin \omega t_0)/\omega$, and the function $f(t)$ and its spectrum are shown in Fig. 6.8.

Multiplication by t^n

If $t^n f(t)$ has an absolutely convergent integral in $-\infty < t < \infty$,

$$\mathscr{F}\{t^n f(t)\} = i^n \frac{d^n}{d\omega^n} F(\omega) \tag{6.45}$$

for $n = 0, 1, 2, \ldots$.

Proof:

$$i^n \frac{d^n}{d\omega^n} F(\omega) = i^n \frac{d^n}{d\omega^n} \int_{-\infty}^{\infty} f(t) e^{-i\omega t}\, dt$$

$$= i^n \int_{-\infty}^{\infty} f(t)(-it)^n e^{-i\omega t}\, dt$$

$$= \mathscr{F}\{t^n f(t)\}. \quad \square$$

For obvious reasons, this property is often referred to as *frequency differentiation*.

Example. Determine $\mathscr{F}\{t^n e^{-t}u(t)\}$.
From Eq. (6.32)

$$\mathscr{F}\{e^{-t}u(t)\} = \frac{1}{i\omega + 1}$$

and therefore

$$\mathscr{F}\{t^n e^{-t}u(t)\} = i^n \frac{d^n}{d\omega^n}\left(\frac{1}{i\omega + 1}\right) = \frac{n!}{(i\omega + 1)^{n+1}};$$

that is,

$$t^n e^{-t}u(t) \Leftrightarrow \frac{n!}{(i\omega + 1)^{n+1}} \tag{6.46}$$

for $n = 0, 1, 2, \ldots$.

Integration

If

$$g(t) = \int_{-\infty}^{t} f(t')\,dt' = D_{-\infty}^{-1} f(t),$$

where $f(t)$ is piecewise-continuous and $f(t)$, $g(t)$ have absolutely convergent integrals in $-\infty < t < \infty$,

$$\mathscr{F}\{D_{-\infty}^{-1} f(t)\} = \frac{F(\omega)}{i\omega}. \tag{6.47}$$

Proof:

$$\begin{aligned}\mathscr{F}\{D_{-\infty}^{-1} f(t)\} &= \int_{-\infty}^{\infty} g(t) e^{-i\omega t}\,dt \\ &= \left[g(t)\frac{e^{-i\omega t}}{-i\omega}\right]_{-\infty}^{\infty} + \frac{1}{i\omega}\int_{-\infty}^{\infty} Dg(t) e^{-i\omega t}\,dt,\end{aligned}$$

using integration by parts. But $g(-\infty) = 0$ from the definition of $g(t)$ and

$$g(\infty) = \int_{-\infty}^{\infty} f(t)\,dt = 0, \tag{6.48}$$

since $g(t)$ has an absolutely convergent integral in $-\infty < t < \infty$. Hence

$$\mathscr{F}\{D_{-\infty}^{-1}f(t)\} = \frac{F(\omega)}{i\omega}$$

and Eq. (6.48) in conjunction with Eq. (6.25) shows that $F(0) = 0$. □

By repeated application of (6.47), we have

$$\mathscr{F}\{D_{-\infty}^{-n}f(t)\} = \frac{F(\omega)}{(i\omega)^n} \tag{6.49}$$

for $n = 0, 1, 2, \ldots$, provided $f, D_{-\infty}^{-1}f, \ldots, D_{-\infty}^{-n}f$ have absolutely convergent integrals in $-\infty < t < \infty$.

Differentiation

If $f(t)$ is continuous and piecewise-smooth and $f(t), Df(t)$ have absolutely convergent integrals in $-\infty < t < \infty$,

$$\mathscr{F}\{Df(t)\} = i\omega F(\omega). \tag{6.50}$$

Proof: Using integration by parts

$$\begin{aligned}\mathscr{F}\{Df(t)\} &= \int_{-\infty}^{\infty} Df(t)e^{-i\omega t}\,dt \\ &= \left[f(t)e^{-i\omega t}\right]_{-\infty}^{\infty} + i\omega\int_{-\infty}^{\infty} f(t)e^{-i\omega t}\,dt \\ &= i\omega F(\omega),\end{aligned}$$

since $f(t) \to 0$ as $t \to \pm\infty$. As shown in Section 6.9, Eq. (6.50) can be justified even if $f(t)$ is only piecewise-continuous, thereby allowing $f(t)$ to have finite jump discontinuities. □

By repeated application of Eq. (6.50) we obtain

$$\mathscr{F}\{D^n f(t)\} = (i\omega)^n F(\omega) \tag{6.51}$$

for $n = 0, 1, 2, \ldots$ provided $f, Df, \ldots, D^{n-1}f$ are continuous, $D^{n-1}f$ is piecewise-smooth, and $f, Df, \ldots, D^n f$ have absolutely convergent integrals in $-\infty < t < \infty$. This property is the dual of (6.45), and because of the $i\omega$ factors in the numerator of the transform, differentiation enhances the high-frequency content of a signal. In contrast, integration reduces the high-frequency content, smoothing the signal and decreasing the more rapid fluctuations.

Example. Determine $\mathscr{F}\{Df(t)\}$ with $f(t) = \text{tri}(t)$.

The triangular function $\text{tri}(t)$ is continuous and its derivative is

$$D\,\text{tri}(t) = u(t+1) - 2u(t) + u(t-1).$$

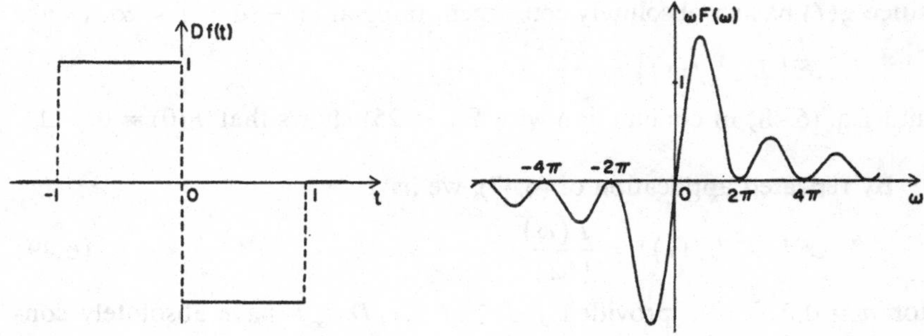

Fig. 6.9: The derivative of tri(t) and its spectrum.

From Eqs. (6.31) and (6.50) the frequency spectrum of $D\,\mathrm{tri}(t)$ is therefore

$$i\omega F(\omega)=4i\frac{\sin^2\omega/2}{\omega},$$

which is purely imaginary, as expected. Since

$$D\,\mathrm{tri}(t)=\mathrm{rect}\left(t+\tfrac{1}{2}\right)-\mathrm{rect}\left(t-\tfrac{1}{2}\right),$$

the spectrum can also be obtained from Eq. (6.14) by using the time-shift property, and the derivative and its spectrum are shown in Fig. 6.9.

Exercises

Using Table 6-2 deduce the Fourier transforms of:

*1. $\cos\omega_0 t\,\mathrm{rect}(2t)$.

2. $\dfrac{\sin 2(t-1)}{2(t-1)}$.

3. $e^{-2t}u(t-1)$.

4. $e^{t}\{1-u(t-1)\}$.

5. $\dfrac{1}{t^2+4}$.

6. $\dfrac{t}{(t^2+1)^2}\sin t$.

Using Table 6-2 deduce the inverse Fourier transforms of:

*7. $\dfrac{1}{(i\omega+1)(i\omega+3)}$.

8. $\dfrac{i\omega}{\omega^2-3i\omega-2}$.

9. $\dfrac{i\omega+1}{\omega^2-2i\omega-2}$.

10. $\dfrac{\sin\omega}{i\omega+2}$.

11. $\dfrac{2}{(i\omega+1)(\omega^2-2i\omega-2)}$.

12. $\dfrac{e^{-2i\omega}}{(i\omega+1)^3}$.

13. $e^{-|\omega|}(3\omega-2)$.

14. $e^{i\omega+\omega}\{1-u(\omega)\}$.

Table 6-2. *Fourier transform relations*

	Time domain	Frequency domain	Source
General			
(i)	$f(t)$	$F(\omega)$	Definition
(ii)	$f(-t)$	$F(-\omega)$	Eq. (6.15)
(iii)	$F(t)$	$2\pi f(-\omega)$	Eq. (6.16)
(iv)	$F(-t)$	$2\pi f(\omega)$	Eq. (6.17)
(v)	$f(at)$	$\frac{1}{\|a\|}F(\frac{\omega}{a})$	Eq. (6.42)
(vi)	$e^{iat}f(t)$	$F(\omega - a)$	Eq. (6.43)
(vii)	$f(t-a)$	$e^{-i\omega a}F(\omega)$	Eq. (6.44)
(viii)	$t^n f(t)$	$i^n \frac{d^n}{d\omega^n}F(\omega)$	Eq. (6.45)
(ix)	$D_{-\infty}^{-n}f(t)$	$\frac{F(\omega)}{(i\omega)^n}$	Eq. (6.49)
(x)	$D^n f(t)$	$(i\omega)^n F(\omega)$	Eq. (6.51)
Particular			
(xi)	rect(t)	$\frac{\sin \omega/2}{\omega/2}$	Eq. (6.14)
(xii)	tri(t)	$\left(\frac{\sin \omega/2}{\omega/2}\right)^2$	Eq. (6.31)
(xiii)	$e^{-t}u(t)$	$\frac{1}{i\omega + 1}$	Eq. (6.32)
(xiv)	$\frac{1}{1 - it}$	$2\pi e^{-\omega}u(\omega)$	(iv) and (xiii)
(xv)	$t^n e^{-t}u(t)$	$\frac{n!}{(i\omega + 1)^{n+1}}$	Eq. (6.46)
(xvi)	$\frac{1}{(1 - it)^{n+1}}$	$\frac{2\pi}{n!}\omega^n e^{-\omega}u(\omega)$	(iv) and (xv)

6.7 Steady-state response

When the results of the previous section are compared with those in Section 2.5, it is evident that the Fourier and Laplace transforms have much in common. There are some problems that can be treated with either transform and others for which only one of them is applicable, but in those instances in which both transforms can be used, the properties suggest that the Fourier transform differs from the Laplace merely in having $i\omega$ in place of s. There are also similarities between the Fourier-transform and Fourier-series techniques. The Fourier-integral representation is analogous to the Fourier-series representation of a periodic function, and the restrictions on the function $f(t)$ are comparable in the two cases. In Chapter 4 we showed that a Fourier series provides an effective tool for finding the steady-state response of a stable system when excited by a periodic input, and it is natural to expect that a Fourier integral will play the same role when $f(t)$ is nonperiodic. Indeed, this was the motivation for considering Fourier transforms at this time.

The steady-state response is an attribute of a stable system only. From the definition given in Section 3.10, if a simple harmonic oscillation $e^{i\omega t}$ is the input to a stable system whose transfer function is $H(s)$, the steady-state response is

$$x(t) = H(i\omega)e^{i\omega t}. \tag{6.52}$$

If, instead, the input is the periodic function $f(t)$ whose complex Fourier-series representation is

$$f(t) = \sum_{n=-\infty}^{\infty} c_n e^{in\omega_0 t}, \tag{6.53}$$

the steady-state response is

$$x(t) = \sum_{n=-\infty}^{\infty} c_n H(in\omega_0) e^{in\omega_0 t} \tag{6.54}$$

as shown in Section 4.7, and from the analogy between Fourier series and integrals, it is evident that if $f(t)$ is a nonperiodic function whose Fourier-integral representation is

$$f(t) = \frac{1}{2\pi}\int_{-\infty}^{\infty} F(\omega)e^{i\omega t}\,d\omega, \tag{6.55}$$

the steady-state response is

$$x(t) = \frac{1}{2\pi}\int_{-\infty}^{\infty} F(\omega)H(i\omega)e^{i\omega t}\,d\omega,$$

that is,

$$x(t) = \mathscr{F}^{-1}\{F(\omega)H(i\omega)\}. \tag{6.56}$$

This follows immediately from Eqs. (6.52) and (6.55) by superposition, and is physically reasonable. Since $H(i\omega)$ is, in effect, the "frequency spectrum of the system," $F(\omega)H(i\omega)$ is the frequency spectrum of the desired response at the output terminals, and the output itself is therefore given by Eq. (6.56).

For a nonperiodic $f(t)$ it is necessary that $f(t) \to 0$ as $t \to \infty$, and the meaning of "steady state" is not as clear as in the periodic case. To better understand what the output really is, consider the solution of the corresponding differential equation. For an nth-order system with $n > 0$

the differential equation is

$$(a_0D^n + a_1D^{n-1} + \cdots + a_n)x(t) = f(t). \tag{6.57}$$

If $f(t)$ is piecewise-continuous, any solution is such that $x(t)$, $Dx(t), \ldots, D^{n-1}x(t)$ are continuous for all t, and if $f(t)$ and $x(t)$ have absolutely convergent integrals in $-\infty < t < \infty$, implying $f(t), x(t) \to 0$ as $t \to \pm\infty$, application of a Fourier transform to Eq. (6.55) gives

$$V(i\omega)X(\omega) = F(\omega),$$

where $V(s)$ is the characteristic function. Hence

$$X(\omega) = F(\omega)H(i\omega) \tag{6.58}$$

and $x(t)$ is the steady-state response given in Eq. (6.56).

The most general solution differs from this by a complementary solution consisting of terms $c_ie^{s_it}$, $i = 1, 2, \ldots, n$, where the s_i are the characteristic roots and the c_i are constants. For a stable system all roots have negative real parts, and as $t \to -\infty$, each term in the complementary solution becomes infinite. No such term can be present in the solution (6.56), and the steady-state response is the only output that is bounded, that is, finite, for all t. It is, in fact, the unique solution of Eq. (6.57), which is zero when $t = -\infty$, and if the input $f(t)$ is zero for $t < 0$, the output is zero for $t < 0$ and corresponds to the system initially at rest (i.e., all initial conditions are zero).

For many systems applications the steady-state response is the only output of practical interest. As we shall now show, it can be obtained without using Fourier transforms at all.

6.8 Convolution

The convolution of two functions $p(t)$ and $q(t)$ existing for all t, $-\infty < t < \infty$, was defined in Section 3.6 as

$$p(t) * q(t) = \int_{-\infty}^{\infty} p(t')q(t-t')\,dt'. \tag{6.59}$$

It is symmetric in the two functions, that is,

$$p(t) * q(t) = q(t) * p(t),$$

and the formation of a convolution involves the four operations of folding and displacement of one function relative to the other, followed by multiplication and then integration. The steps were illustrated in Section 3.6 and, for most functions $p(t)$ and $q(t)$, the direct evaluation of a convolution is a tedious and time-consuming process. It can be simplified using Fourier transforms.

Convolution theorem

Convolution in the time domain is equivalent to multiplication in the frequency domain, and this important property is known as the convolution theorem. The proof is almost identical to that given in Section 3.6 in connection with Laplace transforms.

$$\begin{aligned}\mathscr{F}\{p(t)*q(t)\} &= \int_{-\infty}^{\infty}\int_{-\infty}^{\infty} p(t')q(t-t')e^{-i\omega t}\,dt'\,dt \\ &= \int_{-\infty}^{\infty} p(t')\left(\int_{-\infty}^{\infty} q(t-t')e^{-i\omega t}\,dt\right)dt' \\ &= \int_{-\infty}^{\infty} p(t')\left(\int_{-\infty}^{\infty} q(t)e^{-i\omega t}\,dt\,e^{-i\omega t'}\right)dt' \\ &= \int_{-\infty}^{\infty} p(t')e^{-i\omega t'}\,dt'\int_{-\infty}^{\infty} q(t)e^{-i\omega t}\,dt \\ &= \mathscr{F}\{p(t)\}\mathscr{F}\{q(t)\}.\end{aligned}$$

Thus

$$p(t)*q(t) \Leftrightarrow P(\omega)Q(\omega). \tag{6.60}$$

As shown in the book by W. Kaplan, the theorem is valid if $p(t)$ and $q(t)$ are piecewise-continuous with absolutely convergent integrals in $-\infty < t < \infty$, and if at least one of the two functions is bounded for all t. The convolution is then defined and continuous with absolutely convergent integral in $-\infty < t < \infty$.

Frequency convolution theorem

The convolution theorem has a dual that is known as the frequency convolution theorem and expresses the fact that multiplication in the time domain is equivalent to convolution in the frequency domain. The result can be deduced from (6.60) by using the duality property, but it can also be obtained directly as follows:

$$\begin{aligned}\mathscr{F}\{p(t)q(t)\} &= \int_{-\infty}^{\infty} p(t)q(t)e^{-i\omega t}\,dt \\ &= \int_{-\infty}^{\infty} p(t)\left(\frac{1}{2\pi}\int_{-\infty}^{\infty} Q(\omega')e^{i\omega' t}\,d\omega'\right)e^{-i\omega t}\,dt \\ &= \frac{1}{2\pi}\int_{-\infty}^{\infty} Q(\omega')\left(\int_{-\infty}^{\infty} p(t)e^{-i(\omega-\omega')t}\,dt\right)d\omega' \\ &= \frac{1}{2\pi}\int_{-\infty}^{\infty} Q(\omega')P(\omega-\omega')\,d\omega' \\ &= \frac{1}{2\pi}P(\omega)*Q(\omega).\end{aligned}$$

Thus

$$p(t)q(t) \Leftrightarrow \frac{1}{2\pi}P(\omega) * Q(\omega). \tag{6.61}$$

The theorem provides a simple proof of Parseval's formula (see Section 6.4). If $p(t) = f(t)$ and $q(t) = \bar{f}(t)$ where the bar denotes the complex conjugate,

$$Q(\omega) = \int_{-\infty}^{\infty} \bar{f}(t)e^{-i\omega t}\,dt = \bar{F}(-\omega)$$

and the theorem gives

$$\int_{-\infty}^{\infty} \bar{f}(t)f(t)e^{-i\omega t}\,dt = \frac{1}{2\pi}\int_{-\infty}^{\infty} F(\omega')\bar{F}(\omega' - \omega)\,d\omega'.$$

On putting $\omega = 0$ we obtain

$$\int_{-\infty}^{\infty} |f(t)|^2\,dt = \frac{1}{2\pi}\int_{-\infty}^{\infty} |F(\omega')|^2\,d\omega'$$

in agreement with Eq. (6.28).

Correlation

The process of correlating two signals $p(t)$ and $q(t)$ is mathematically similar to forming their convolution. The correlation function is defined as

$$\begin{aligned} r(t) &= \int_{-\infty}^{\infty} p(t')\bar{q}(t' - t)\,dt' \\ &= \int_{-\infty}^{\infty} p(t' + t)\bar{q}(t')\,dt', \end{aligned} \tag{6.62}$$

where the bar denotes the complex conjugate, and clearly

$$r(t) = p(t) * \bar{q}(-t). \tag{6.63}$$

In general it is not symmetric in p and q.

The significant differences between correlation and convolution are that the complex conjugate of one function is involved and that neither function is folded on the other. The correlation function is a measure of the similarity of $p(t)$ and $q(t)$ as a function of the displacement of $q(t)$ to the right or of $p(t)$ to the left. If, for example, a seismic signal is received at two locations, the actual waveforms may look quite different because of their different propagation paths, but by computing the correlation function, the value of t at which $r(t)$ is a maximum specifies the time lag between the two receivers. When two different signals are correlated, $r(t)$ is referred to as the *cross-correlation* function, whereas if $q(t) = p(t)$, $r(t)$ is called the *autocorrelation* function. The latter is a measure of the repetitive character of a signal.

As evident from Eq. (6.63),

$$\mathscr{F}\{r(t)\} = R(\omega) = \mathscr{F}\{p(t)\}\mathscr{F}\{\bar{q}(-t)\}$$

and since

$$\mathscr{F}\{\bar{q}(-t)\} = \int_{-\infty}^{\infty} \bar{q}(-t)e^{-i\omega t}\,dt = \overline{Q}(\omega),$$

$$R(\omega) = P(\omega)\overline{Q}(\omega). \tag{6.64}$$

In the particular case of autocorrelation,

$$R(\omega) = |P(\omega)|^2 \tag{6.65}$$

and $R(\omega)/(2\pi)$ is the energy spectral density. Parseval's formula now follows immediately from Eqs. (6.62) and (6.65).

If $q(t)$ is a real function so that $\bar{q}(t) = q(t)$, then

$$R(\omega) = P(\omega)Q(-\omega),$$

and if, in addition, $q(t)$ is even, so that $Q(-\omega) = Q(\omega)$, we have

$$R(\omega) = P(\omega)Q(\omega).$$

As otherwise evident from Eq. (6.62), correlation and convolution are now identical operations.

The correlation function is just as difficult to evaluate as a convolution, and there are only a few cases where it is easy to do directly.

Example. Determine the correlation function for the two rectangular pulses

$$p(t) = 2\{u(t-4) - u(t-6)\} \quad \text{and} \quad q(t) = u(t) - u(t-3).$$

The functions are shown in Fig. 6.10(a), and when substituted into Eq. (6.62) we have

$$r(t) = 2\int_4^6 \{u(t'-t) - u(t'-3-t)\}\,dt'.$$

This can be evaluated by considering different ranges of t separately.

If $t + 3 < 4$, i.e., $t < 1$,	$r(t) = 0$;
if $4 < t + 3 < 6$, i.e., $1 < t < 3$,	$r(t) = 2\int_4^{t+3} dt' = 2(t-1)$;
if $t < 4$ but $6 < t + 3$, i.e., $3 < t < 4$,	$r(t) = 2\int_4^6 dt' = 4$;
if $4 < t < 6$,	$r(t) = 2\int_t^6 dt' = 2(6-t)$;
if $6 < t$,	$r(t) = 0$.

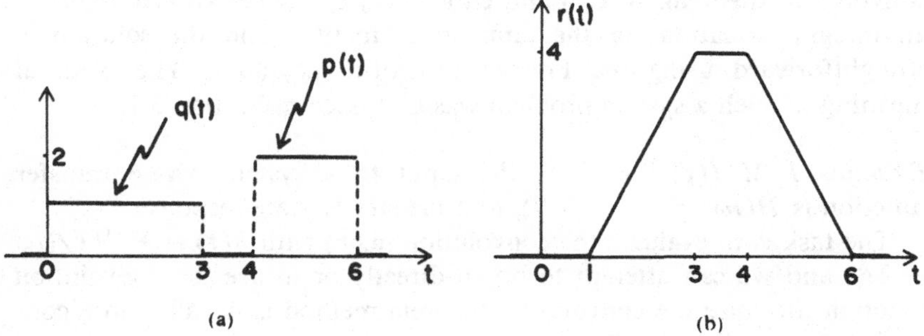

Fig. 6.10: The functions $p(t)$ and $q(t)$ and their correlation function $r(t)$.

The resulting correlation function $r(t)$ is plotted in Fig. 6.10(b), and we observe that it is a maximum during the time when the longer pulse spans the shorter.

Application to a system

We showed in Section 6.7 that for a stable system the steady-state response produced by an input $f(t)$ is

$$x(t) = \mathscr{F}^{-1}\{F(\omega)H(i\omega)\},$$

which can now be expressed as a convolution:

$$x(t) = h(t) * f(t), \tag{6.66}$$

where

$$h(t) = \mathscr{F}^{-1}\{H(i\omega)\}. \tag{6.67}$$

If $h(t)$ is the impulse response of the system (a fact not yet established), the output is equivalent to that obtained in Section 3.7 on the assumption of an input that is zero for $t < 0$ and a system initially at rest. At the moment, (6.66) is valid only for a stable system and an input $f(t)$ that tends to zero as $t \to \pm\infty$, but since we are no longer restricted to inputs that are zero for $t < 0$, the result is, in this respect, a generalization of the previous one.

From Eq. (6.66)

$$x(t) = \int_{-\infty}^{\infty} h(t')f(t - t')\,dt' \tag{6.68}$$

and if $f(t)$ and $h(t)$ or, equivalently, $H(i\omega)$ are given, $x(t)$ can be found by the direct evaluation of the convolution integral or by invoking the convolution theorem. If $x(t)$ and either $h(t)$ or $f(t)$ is known, (6.68) is an integral equation for the remaining function, and the solution is straightforward using the Fourier-transform technique. The physical meaning of such a system problem was discussed in Section 3.7.

Example 1. If $f(t) = e^{-|t|}$ is the input to a system whose transfer function is $H(i\omega) = 1/(i\omega + 2)$, find the steady-state response.

The task is to evaluate the convolution (6.66) with $h(t) = \mathscr{F}^{-1}\{1/(i\omega + 2)\}$, and we can attempt to do so directly or to use the convolution theorem. Because the convolution-theorem method is usually more convenient, we shall use it. Since

$$\begin{aligned} F(\omega) &= \int_{-\infty}^{\infty} e^{-|t|}e^{-i\omega t}\,dt \\ &= \int_{-\infty}^{0} e^{(1-i\omega)t}\,dt + \int_{0}^{\infty} e^{-(1+i\omega)t}\,dt \\ &= \frac{2}{\omega^2+1}, \end{aligned}$$

we have

$$X(\omega) = H(i\omega)F(\omega) = \frac{2}{(i\omega+2)(\omega^2+1)}.$$

Hence

$$\begin{aligned} x(t) &= \frac{1}{\pi}\int_{-\infty}^{\infty} \frac{1}{(i\omega+2)(\omega^2+1)} e^{i\omega t}\,d\omega \\ &= \int_{-\infty}^{\infty} g(z)\,dz, \end{aligned}$$

where

$$g(z) = -\frac{i}{\pi}\frac{e^{izt}}{(z-2i)(z-i)(z+i)}.$$

If $t < 0$ the path closure must be in the LHP, giving

$$x(t) = -2\pi i\,\text{Res}\{g(z), -i\} = \tfrac{1}{3}e^{t},$$

whereas if $t > 0$ the closure must be in the UHP, giving

$$\begin{aligned} x(t) &= 2\pi i\left[\text{Res}\{g(z), 2i\} + \text{Res}\{g(z), i\}\right] \\ &= e^{-t} - \tfrac{2}{3}e^{-2t}. \end{aligned}$$

The steady-state response is therefore

$$x(t) = \tfrac{1}{3}e^{-|t|} + \tfrac{2}{3}e^{-t}(1 - e^{-t})u(t).$$

Example 2. Find $f(t)$ where

$$f(t) - \int_0^{\infty} t' e^{-2t'} f(t - t')\,dt' = 4e^{-t}(\cos 2t)u(t)$$

for all t, $-\infty < t < \infty$.

We observe that the integral is

$$\{te^{-2t}u(t)\} * f(t)$$

and by application of a Fourier transform, we obtain

$$F(\omega) - \mathscr{F}\{te^{-2t}u(t)\}F(\omega) = \mathscr{F}\{4e^{-t}(\cos 2t)u(t)\};$$

that is,

$$F(\omega)\left\{1 - \frac{1}{(i\omega + 2)^2}\right\} = \frac{4(i\omega + 1)}{(i\omega + 1)^2 + 4}.$$

On simplifying, this becomes

$$F(\omega) = \frac{4(i\omega + 2)^2}{(i\omega + 3)\{(i\omega + 1)^2 + 4\}}$$

and hence

$$f(t) = \frac{2}{\pi}\int_{-\infty}^{\infty} \frac{(i\omega + 2)^2}{(i\omega + 3)\{(i\omega + 1)^2 + 4\}} e^{i\omega t}\,d\omega$$

$$= \int_{-\infty}^{\infty} g(z)\,dz,$$

where

$$g(z) = -\frac{2i}{\pi}\frac{(z - 2i)^2 e^{izt}}{(z - 3i)(z - i - 2)(z - i + 2)}.$$

For $t < 0$ path closure in the LHP gives zero, whereas for $t > 0$ closure in the UHP gives

$$f(t) = 2\pi i[\operatorname{Res}\{g(z), 3i\} + \operatorname{Res}\{g(z), i + 2\} + \operatorname{Res}\{g(z), i - 2\}]$$
$$= \tfrac{1}{2}\{e^{-3t} + e^{-t}(\sin 2t + 7\cos 2t)\}.$$

The solution of the integral equation is therefore

$$f(t) = \tfrac{1}{2}\{e^{-3t} + e^{-t}(\sin 2t + 7\cos 2t)\}u(t).$$

In each example the residue evaluation could have been avoided by using the Fourier-transform relations in Table 6-2, and the reader may wish to verify this.

Exercises

In the following, a, b, and ω_0 are positive real constants.

1. For the signal $f(t) = e^{-at}u(t)$, show that:
   (i) The energy contained in the frequency band $|\omega| < \omega_0$ is
$$E(\omega_0) = \frac{1}{\pi a}\arctan\frac{\omega_0}{a}.$$
   (ii) $\lim_{\omega_0 \to \infty} E(\omega_0) = E$ where E is the energy computed in the time domain.

2. Verify Parseval's formula for the signal
$$f(t) = te^{-at}u(t).$$

3. Verify Parseval's formula for the signal
$$f(t) = u(t) - u(t-a).$$

By direct evaluation find $p(t) * q(t)$ for the following functions.

4. $e^{-at}u(t)$, $e^{-bt}u(t)$ with $b \neq a$.
5. $e^{-at}u(t)$, $e^{-at}u(t)$.
6. $te^{-t}u(t)$, $e^{-t}u(t)$.
7. $te^{-t}u(t)$, $e^{t}\{1 - u(t)\}$.
8. $e^{-at}u(t)$, $(1-t)\{u(t) - u(t-1)\}$.
9. $(\sin 2\pi t)\{u(t) - u(t-\frac{1}{8})\}$, $u(t) - u(t-\frac{1}{2})$.
10. (i) By direct evaluation of the convolution integral, show that
$$\text{rect}(t) * \text{rect}(t) = \text{tri}(t),$$
   where rect(t) and tri(t) are the functions defined in Eqs. (6.13) and (6.30), respectively.
   (ii) Deduce that
$$\int_{-\infty}^{\infty}\left(\frac{\sin\omega}{\omega}\right)^2 d\omega = \pi.$$

(iii) From the convolution of tri(t) with rect(t) and tri(t), show that

$$\int_{-\infty}^{\infty}\left(\frac{\sin\omega}{\omega}\right)^3 d\omega=\frac{3\pi}{4}\quad\text{and}\int_{-\infty}^{\infty}\left(\frac{\sin\omega}{\omega}\right)^4 d\omega=\frac{2\pi}{3}.$$

11. Determine the "width" of $(1+t^2/a^2)^{-1}*(1+t^2/b^2)^{-1}$ in terms of the widths of the individual functions.

12. Show that

$$\frac{1}{\pi}\int_{-\infty}^{\infty}\frac{\sin at'}{t'}\frac{\sin(t-t')}{t-t'}\,dt'=\begin{cases}\dfrac{\sin t}{t} & \text{if } a>1,\\[2ex] \dfrac{\sin at}{t} & \text{if } a<1.\end{cases}$$

If $f(t)+\frac{3}{2}\int_{-\infty}^{\infty}e^{-|t'|}f(t-t')\,dt'=g(t)$ for all t, $-\infty<t<\infty$, find $f(t)$ for the following $g(t)$.

*13. $g(t)=e^{-|t|}$.

14. $g(t)=te^{-2|t|}$.

6.9 The impulse and other functions

Until now we have restricted attention to functions that satisfy the conditions given in Section 6.3 for the existence of a Fourier transform. This has forced us to exclude such simple functions as $\sin t$, $u(t)$, and, most particularly, $\delta(t)$, none of which has a transform in the classical sense, but since the conditions were sufficient and not necessary, it is feasible that the theory could be extended to include these other functions. It is important that this be done. In Section 6.8 the fact that $\mathscr{F}\{\delta(t)\}$ was not defined made it impossible to describe $\mathscr{F}^{-1}\{H(i\omega)\}$; and more to the point, the full power of the Fourier-transform technique for the analysis of signals and systems cannot be achieved without the inclusion of generalized functions.

Fourier transform of $\delta(t)$

The impulse function and its properties were described in Section 3.3, and our objective now is to define the Fourier transform of $\delta(t)$ in a manner consistent with these properties and those of the basic Fourier-transform operation.

From the sifting property of $\delta(t)$ given in Eq. (3.12), we are led to the definition

$$\mathscr{F}\{\delta(t)\} = \int_{-\infty}^{\infty} \delta(t) e^{-i\omega t}\, dt = 1,$$

and from the definition of a transform pair it is therefore necessary that

$$\delta(t) = \mathscr{F}^{-1}\{1\} = \frac{1}{2\pi}\int_{-\infty}^{\infty} e^{i\omega t}\, d\omega.$$

Thus

$$\delta(t) \Leftrightarrow 1, \tag{6.69}$$

which confirms that $\delta(t)$ is an even function, and Eq. (6.16) now implies

$$1 \Leftrightarrow 2\pi\delta(\omega). \tag{6.70}$$

Since a constant does not have a direct or inverse Fourier transform in classical theory, these new definitions produce no inconsistencies. Similarly

$$\mathscr{F}\{\delta(t - t_0)\} = \int_{-\infty}^{\infty} \delta(t - t_0) e^{-i\omega t}\, dt = e^{-i\omega t_0},$$

which is in agreement with the time-shift property of a Fourier transform applied to (6.69), and therefore

$$\delta(t - t_0) \Leftrightarrow e^{-i\omega t_0}. \tag{6.71}$$

The dual of this is

$$e^{i\omega_0 t} \Leftrightarrow 2\pi\delta(\omega - \omega_0), \tag{6.72}$$

which is consistent with the frequency-shift property applied to (6.70).

The preceding four transform pairs are shown in Fig. 6.11; none of them exists in classical theory. Each can be justified using the theory of distributions (see Appendix 1 of the book by A. Papoulis for a summary of the relevant aspects), but from a practical point of view it is sufficient if the procedures for interpreting them are internally consistent and compatible with classical theory. In this connection we note that the normalization of $\delta(t)$ [see Eq. (3.11)] implies

$$\delta(at) = \frac{1}{|a|}\delta\left(\frac{t}{a}\right), \tag{6.73}$$

and that this is in accordance with the scaling property (6.42).

The Fourier-integral representation of the impulse function provides a simple, albeit formal, proof of Fourier's identity (6.5). From Eq. (6.71)

$$\int_{-\infty}^{\infty} e^{i\omega(t-t')}\, d\omega = 2\pi\delta(t - t')$$

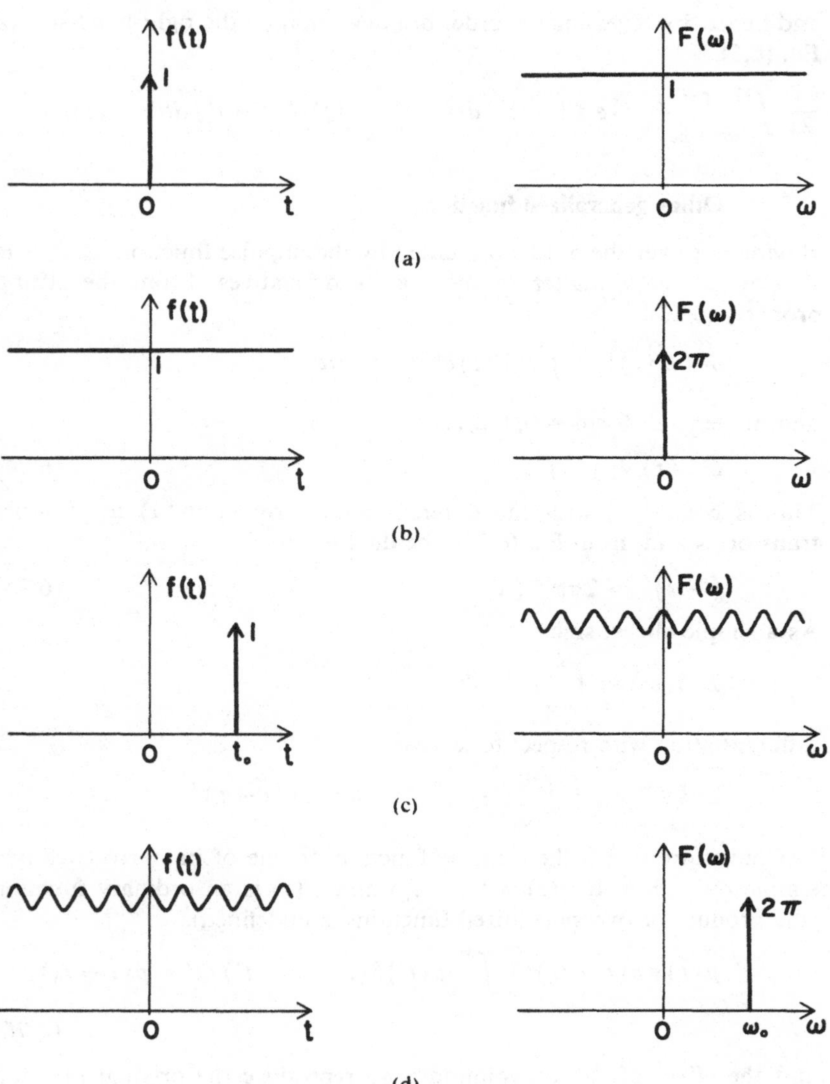

Fig. 6.11: The Fourier-transform pairs of Eqs. (6.69) through (6.72). The wiggly lines in (c) and (d) represent the exponentials $e^{-i\omega t_0}$ and $e^{i\omega_0 t}$, respectively.

and hence, by reversing the order of integration on the right-hand side of Eq. (6.5),

$$\frac{1}{2\pi}\int_{-\infty}^{\infty}\int_{-\infty}^{\infty} f(t')e^{i\omega(t-t')}\,dt'\,d\omega = \int_{-\infty}^{\infty} f(t')\delta(t-t')\,dt' = f(t).$$

Other generalized functions

Having got over the hurdle presented by the impulse function itself, it is a relatively easy matter to include its derivatives. From the sifting property (3.13),

$$\mathscr{F}\{\delta'(t)\} = \int_{-\infty}^{\infty} \delta'(t)e^{-i\omega t}\,dt = i\omega$$

and, in general, for $n = 0, 1, 2, \ldots,$

$$\delta^{(n)}(t) \Leftrightarrow (i\omega)^n. \tag{6.74}$$

This is consistent with the differentiation formula (6.51) for Fourier transforms and, from Eq. (6.17), the dual is

$$(-it)^n \Leftrightarrow 2\pi\delta^{(n)}(\omega). \tag{6.75}$$

As a simple check, since

$$2\pi\delta(\omega) = \int_{-\infty}^{\infty} e^{-i\omega t}\,dt,$$

differentiation with respect to ω gives

$$2\pi\delta^{(n)}(\omega) = \int_{-\infty}^{\infty} (-it)^n e^{-i\omega t}\,dt = \mathscr{F}\{(-it)^n\}.$$

Convolution with the impulse function or one of its derivatives is a simple operation. If $q(t) = \delta(t - t_0)$ and $p(t)$ is an ordinary function (the product of two generalized functions is undefined),

$$p(t) * \delta(t - t_0) = \int_{-\infty}^{\infty} p(t')\delta(t - t_0 - t')\,dt' = p(t - t_0), \tag{6.76}$$

and the effect of the convolution is to reproduce the original function $p(t)$, shifted according to the displacement of the impulse. Similarly, from Eq. (3.13),

$$p(t) * \delta^{(n)}(t - t_0) = p^{(n)}(t - t_0) \tag{6.77}$$

and the shift is now accompanied by differentiation.

The introduction of generalized functions also makes it possible to relax the continuity requirement for the validity of the differentiation formula (6.50) by allowing $f(t)$ to have a finite number of jump discontinuities; $f'(t)$ is then a generalized function. As an example, consider

$$f(t) = g(t)u(t),$$

where $g(t)$ is assumed to have a continuous derivative for all t. Since $Du(t) = \delta(t)$,

$$f'(t) = g'(t)u(t) + g(0)\delta(t)$$

and

$$\begin{aligned}\mathscr{F}\{g'(t)u(t)\} &= \int_0^\infty g'(t)e^{-i\omega t}\,dt \\ &= \left[g(t)e^{-i\omega t}\right]_0^\infty + i\omega\int_0^\infty g(t)e^{-i\omega t}\,dt \\ &= -g(0) + i\omega F(\omega).\end{aligned}$$

Hence

$$\begin{aligned}\mathscr{F}\{f'(t)\} &= -g(0) + i\omega F(\omega) + \mathscr{F}\{g(0)\delta(t)\} \\ &= i\omega F(\omega),\end{aligned}$$

in agreement with Eq. (6.50) in spite of the possible jump discontinuity in $f(t)$ at $t = 0$.

The transform pair (6.71) in combination with the integration formula for Fourier transforms provides a simple method for finding the transform of a piecewise-linear pulse whose first or second derivative is a collection of impulse functions. An example is the triangular function $f(t) = \operatorname{tri}(t)$. We have

$$D\operatorname{tri}(t) = u(t+1) - 2u(t) + u(t-1)$$

and

$$D^2\operatorname{tri}(t) = \delta(t+1) - 2\delta(t) + \delta(t-1),$$

and the function and its first two derivatives are shown in Fig. 6.12. From (6.71)

$$\begin{aligned}\mathscr{F}\{D^2\operatorname{tri}(t)\} &= e^{i\omega} - 2 + e^{-i\omega} \\ &= 2(\cos\omega - 1) = -4\sin^2\frac{\omega}{2},\end{aligned}$$

and the integration formula (6.49) now gives

$$\mathscr{F}\{\operatorname{tri}(t)\} = -4\frac{\sin^2\omega/2}{(i\omega)^2} = \left(\frac{\sin\omega/2}{\omega/2}\right)^2$$

in accordance with (6.31).

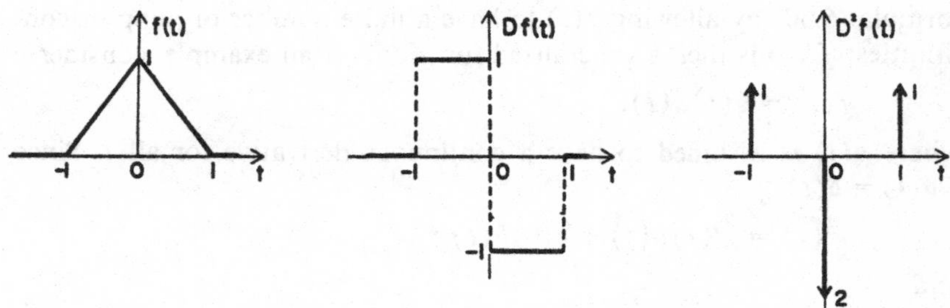

Fig. 6.12: The triangular pulse $f(t) = \text{tri}(t)$ and its first two derivatives.

The method is valid only for a function $f(t)$ whose transform exists in the classical sense. Since $Du(t) = \delta(t)$, it follows that

$$\mathscr{F}\{Du(t)\} = 1, \tag{6.78}$$

from which it is tempting to conclude

$$\mathscr{F}\{u(t)\} = \frac{1}{i\omega}. \tag{6.79}$$

This is inconsistent with the known properties of $u(t)$, as evident from the identity

$$u(t) + u(-t) = 1, \qquad t \neq 0,$$

which implies

$$\mathscr{F}\{u(t)\} + \mathscr{F}\{u(-t)\} = \mathscr{F}\{1\} = 2\pi\delta(\omega). \tag{6.80}$$

However, an alternative deduction from Eq. (6.78) is

$$\mathscr{F}\{u(t)\} = \frac{1}{i\omega} + a\delta(\omega),$$

where a is some constant, and this satisfies Eq. (6.72) if $a = \pi$. We therefore take

$$u(t) \Leftrightarrow \frac{1}{i\omega} + \pi\delta(\omega) \tag{6.81}$$

and this should be regarded as a definition. The transform is peculiar in that $1/i\omega$ is an ordinary function, discontinuous at $\omega = 0$. The unit-step function $u(t)$ is likewise an ordinary function, but since it remains finite as $t \to \infty$, it does not have a transform in the classical sense. The result illustrates the fact that if generalized functions are included, the equation $\omega F_1(\omega) = \omega F_2(\omega)$ does *not* imply $F_1(\omega) = F_2(\omega)$.

The Fourier-integral representation of $u(t)$ is

$$u(t) = \frac{1}{2\pi}\int_{-\infty}^{\infty}\left\{\frac{1}{i\omega} + \pi\delta(\omega)\right\}e^{i\omega t}\,d\omega$$

and thus

$$\begin{aligned} u(t) &= \frac{1}{2} + \frac{1}{2\pi i}\int_{-\infty}^{\infty}\frac{1}{\omega}e^{i\omega t}\,d\omega \\ &= \frac{1}{2} + \frac{1}{2\pi}\int_{-\infty}^{\infty}\frac{\sin\omega t}{\omega}\,d\omega, \end{aligned}$$

where we have used the property

$$\int_{-\infty}^{\infty} f(x)\,dx = 0$$

of an odd function. The integral can be evaluated by path closure (see Example 3 in Section 5.7) and the result is

$$u(t) = \begin{cases} 0 & \text{for } t < 0, \\ \frac{1}{2} & \text{for } t = 0, \\ 1 & \text{for } t > 0. \end{cases}$$

As expected, the transform defines $u(t)$ even at its point of jump discontinuity.

The presence of the impulse function in the definition of $\mathscr{F}\{u(t)\}$ is unfortunate. It is a source of error and confusion, and the fact that the Laplace transform does not have it is one reason why problems for which both transforms are applicable are more often treated using Laplace-transform techniques. In the context of Fourier transforms, the unit-step function is a pathological case. If $\sigma > 0$

$$\mathscr{F}\{e^{-\sigma t}u(t)\} = \frac{1}{i\omega + \sigma},$$

which is an ordinary, well-behaved function. Moreover, if $f(t)$ is an ordinary function that includes $u(t)$ but satisfies the requirement that $f(t) \to 0$ as $t \to \infty$, the impulse function has no effect on the transform and can be ignored. As an example

$$\begin{aligned} \mathscr{F}\{\mathrm{rect}(t)\} &= \mathscr{F}\left\{u\left(t + \frac{1}{2}\right)\right\} - \mathscr{F}\left\{u\left(t - \frac{1}{2}\right)\right\} \\ &= e^{i\omega/2}\left\{\frac{1}{i\omega} + \pi\delta(\omega)\right\} - e^{-i\omega/2}\left\{\frac{1}{i\omega} + \pi\delta(\omega)\right\} \\ &= \frac{1}{i\omega}e^{i\omega/2} + \pi\delta(\omega) - \frac{1}{i\omega}e^{-i\omega/2} - \pi\delta(\omega) \\ &= \frac{1}{i\omega}\left(e^{i\omega/2} - e^{-i\omega/2}\right), \end{aligned}$$

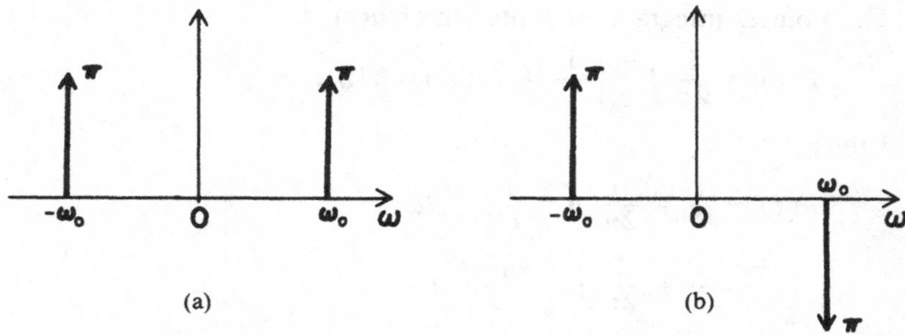

Fig. 6.13: The frequency spectra of (a) $\cos\omega_0 t$ and (b) $-i\sin\omega_0 t$.

which is the same as if we had used the simpler but incorrect transform (6.79). This happens with most problems in signal analysis and explains why it is not uncommon for those who are immersed in this subject to have developed devious techniques that, when used consistently, "eliminate the need" for the impulse function in (6.81).

Periodic functions

A periodic function does not have a Fourier transform in the classical sense, but with the introduction of the impulse function it is now possible to give a meaning to the transform.

The simplest of all periodic functions are the cosine and sine, and since

$$\cos\omega_0 t = \tfrac{1}{2}\left(e^{i\omega_0 t} + e^{-i\omega_0 t}\right),$$

if follows from Eq. (6.72) that

$$\cos\omega_0 t \Leftrightarrow \pi\{\delta(\omega+\omega_0) + \delta(\omega-\omega_0)\}$$

and similarly

$$\sin\omega_0 t \Leftrightarrow i\pi\{\delta(\omega+\omega_0) - \delta(\omega-\omega_0)\}.$$

The spectrum of each is a pair of impulse functions, and these are shown in Fig. 6.13.

For a general periodic function $f(t)$, the frequency spectrum can be obtained, formally at least, by transforming the Fourier-series representation of $f(t)$ term by term. If

$$f(t) = \sum_{n=-\infty}^{\infty} c_n e^{in\omega_0 t}, \qquad (6.82)$$

where $\tau = 2\pi/\omega_0$ is the period of $f(t)$ and the c_n are the Fourier coefficients defined in Eq. (6.2), application of a Fourier transform gives

$$F(\omega) = 2\pi \sum_{n=-\infty}^{\infty} c_n \delta(\omega - n\omega_0). \tag{6.83}$$

Equations (6.82) and (6.83) contain the same information about $f(t)$, and either can be used to plot the frequency spectrum, but if (6.83) is employed, the spectrum consists of an infinite set of impulse functions of strength $2\pi c_n$ at the frequencies $\omega = n\omega_0$ with $n = 0, \pm 1, \pm 2, \ldots$. Since $[F(\omega)]/(2\pi)$ is the spectral density, this picture is consistent with the line spectrum of a periodic function described in Section 4.7.

A more useful form for the Fourier transform is obtained as follows. Let

$$g(t) = f(t)\,\mathrm{rect}\left(\frac{t}{\tau}\right)$$

so that, in the case of a pulse train, $g(t)$ consists of only a single pulse. Then

$$f(t) = \sum_{n=-\infty}^{\infty} g(t - n\tau) \tag{6.84}$$

and

$$\begin{aligned} F(\omega) &= \sum_{n=-\infty}^{\infty} \mathscr{F}\{g(t - n\tau)\} \\ &= \sum_{n=-\infty}^{\infty} e^{-in\omega\tau} G(\omega) \end{aligned} \tag{6.85}$$

using the time-shift property (6.44). This must be equivalent to (6.83), and to show that it is, consider

$$S_N(\omega) = \sum_{n=-N}^{N} e^{-in\omega\tau}.$$

From the formula for the sum of a geometric series, we have

$$S_N(\omega) = \frac{e^{i(N+1)\omega\tau} - e^{-iN\omega\tau}}{e^{i\omega\tau} - 1} = \frac{\sin(N + 1/2)\omega\tau}{\sin(\omega\tau/2)},$$

which is clearly a periodic function of ω. The period is $\omega_0 = 2\pi/\tau$ and the function is shown in Fig. 6.14 for $N = 7$. Peaks of magnitude $2N + 1$ occur at $\omega = n\omega_0$, $n = 0, \pm 1, \pm 2, \ldots$, and between them the function is small and oscillatory. As N increases the oscillations become more rapid and although their amplitudes do not decrease, they become less signifi-

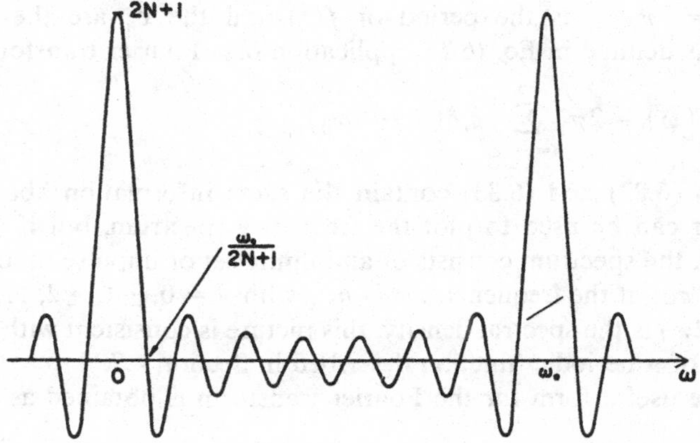

Fig. 6.14: The function $S_7(\omega)$ with $\omega_0 = 2\pi/\tau$.

cant in comparison with the peaks. For large N the area under the peak is approximately ω_0, and, as shown by A. Papoulis (pp. 43–5 of the book), in the limit as $N \to \infty$, $S_N(\omega)$ becomes an infinite set of impulse functions of strength ω_0 spaced ω_0 apart; that is,

$$\lim_{N\to\infty} S_N(\omega) = \omega_0 \sum_{n=-\infty}^{\infty} \delta(\omega - n\omega_0).$$

Hence

$$\sum_{n=-\infty}^{\infty} e^{-in\omega\tau} = \omega_0 \sum_{n=-\infty}^{\infty} \delta(\omega - n\omega_0) \tag{6.86}$$

with $\omega_0 = 2\pi/\tau$, and the frequency spectrum of $f(t)$ is therefore

$$F(\omega) = \omega_0 \sum_{n=-\infty}^{\infty} \delta(\omega - n\omega_0)G(\omega). \tag{6.87}$$

Comparison with Eq. (6.83) shows, formally at least,

$$G(\omega) = \tau c_n,$$

and it will be recalled that this was the basis for the definition of a Fourier transform in Section 6.1.

Equation (6.86) is also the source of a new transform pair, and since

$$\mathscr{F}^{-1}\{e^{-in\omega\tau}\} = \delta(t - n\tau),$$

we have

$$\sum_{n=-\infty}^{\infty} \delta(t - n\tau) \Leftrightarrow \omega_0 \sum_{n=-\infty}^{\infty} \delta(\omega - n\omega_0). \tag{6.88}$$

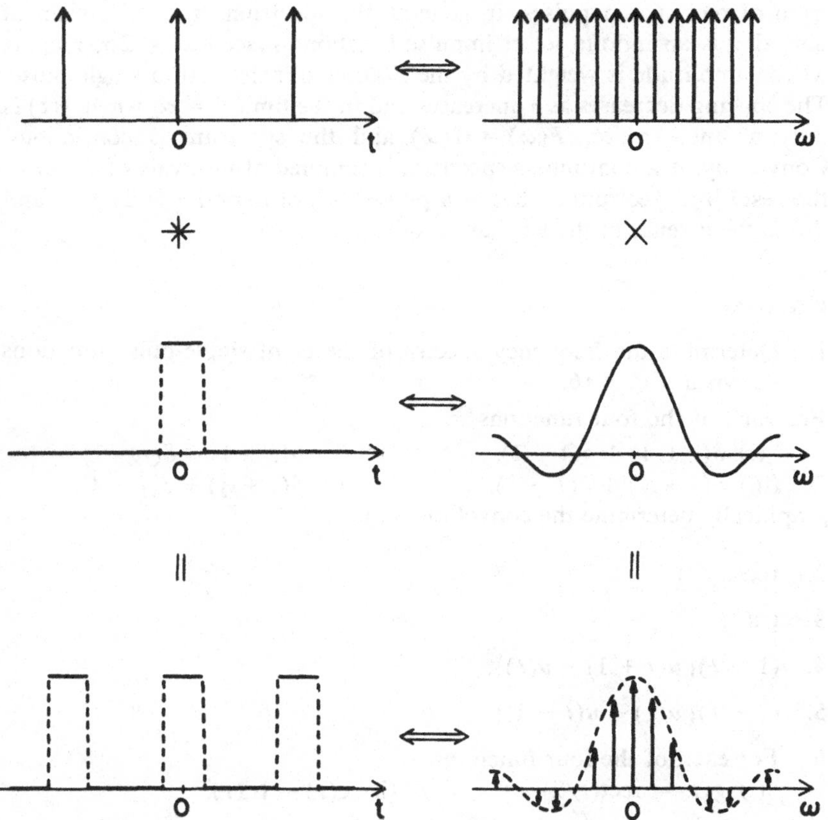

Fig. 6.15: The construction of a periodic function $f(t)$ and the determination of its frequency spectrum.

Using this it is possible to determine the frequency spectrum of a periodic function directly, without reference to Fourier series. From Eq. (6.76) the periodic function $f(t)$ given in Eq. (6.84) can be written as

$$f(t) = \sum_{n=-\infty}^{\infty} \delta(t - n\tau) * g(t). \tag{6.89}$$

By the convolution theorem its frequency spectrum is

$$F(\omega) = \mathscr{F}\left\{ \sum_{n=-\infty}^{\infty} \delta(t - n\tau) \right\} \mathscr{F}\{ g(t) \},$$

which is simply Eq. (6.87). The process is illustrated in Fig. 6.15 for a train of rectangular pulses. In general, the spectrum of a pulse train of period τ is an infinite set of impulse functions spaced $\omega_0 = 2\pi/\tau$ apart whose amplitude is weighted by the Fourier transform of a single pulse. The spacing decreases as τ increases and in the limit $\tau = \infty$ when $f(t)$ is only a single pulse, $F(\omega) = G(\omega)$ and the spectrum is continuous. Conversely, if a continuous spectrum is sampled at intervals of ω_0 in ω, the resulting spectrum is that of a pulse train of period $\tau = 2\pi/\omega_0$, and this is the essence of the FFT technique.

Exercises

1. Determine the frequency spectra of the eight single-pulse functions shown in Fig. 6.16.

For each of the four functions

(i) $\delta(t+1) + \delta(t-1)$, (ii) $\delta(t+1) - \delta(t-1)$,
(iii) $\delta(t+\frac{1}{2}) + \delta(t-\frac{1}{2})$, (iv) $\delta(t+\frac{1}{2}) + \delta(t-1)$,

graphically determine the convolution with:

2. $\mathrm{rect}(t/2)$.
3. $\mathrm{tri}(t)$.
4. $(1+t)\{u(t+1) - u(t)\}$.
5. $(1-t)\{u(t) - u(t-1)\}$.
6. For each of the four functions
   (i) $g(t) = \mathrm{rect}(t/2)$, (ii) $g(t) = \mathrm{tri}(t)$,
   (iii) $g(t) = \cos\frac{\pi t}{2}\mathrm{rect}(t/2)$, (iv) $g(t) = \sin \pi t\,\mathrm{rect}(t/2)$,
   plot the periodic function $f(t) = \Sigma_{n=-\infty}^{\infty} g(t - n\tau)$ with $\tau = 4$, and then compute and display the frequency spectrum $F(\omega)$.
7. Repeat Exercise 6 with $\tau = 2$.

6.10 Systems applications

The results of the preceding sections show that the response of a stable system to a rather general input can be determined using Fourier transforms. With the inclusion of generalized functions, it is possible to admit inputs that are piecewise-continuous in $-\infty < t < \infty$ and of less than exponential growth as $t \to -\infty$, for example, the function t^2, or even generalized functions themselves, and the resulting output is

$$x(t) = \mathscr{F}^{-1}\{H(i\omega)F(\omega)\}, \tag{6.90}$$

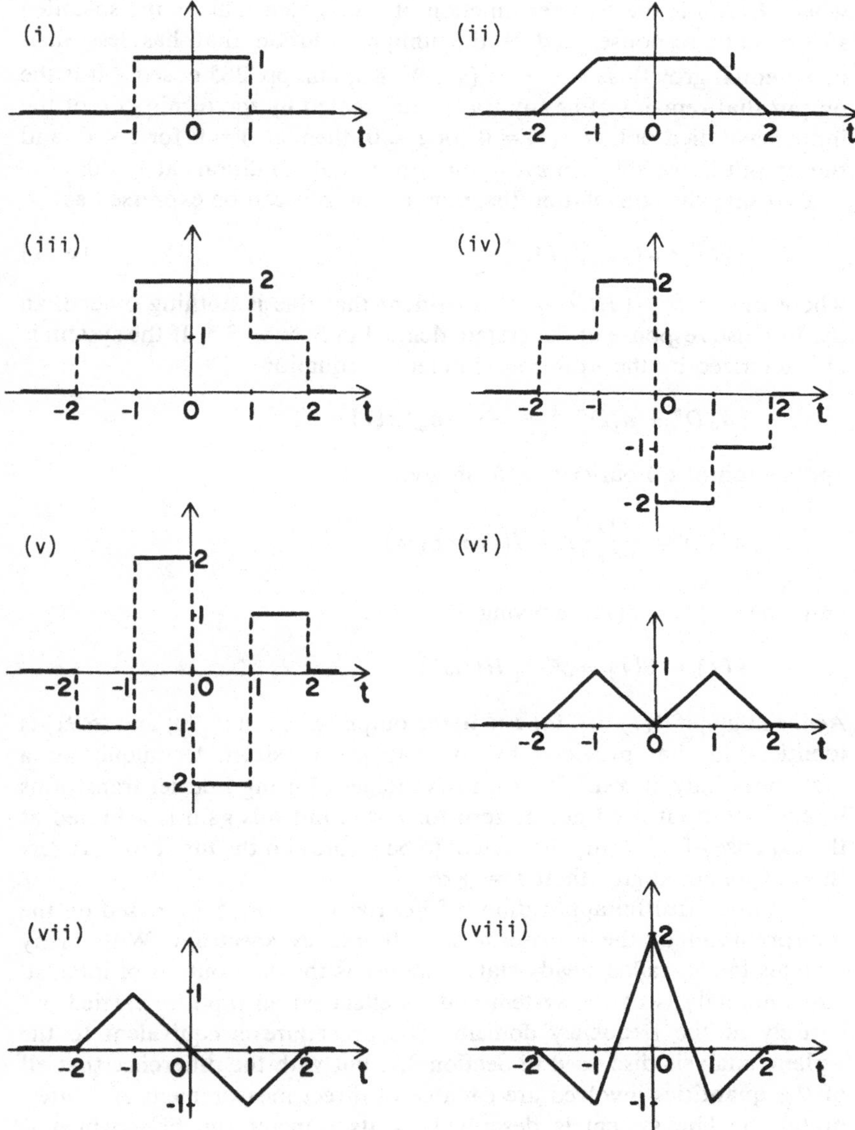

Fig. 6.16: Single-pulse functions $f(t)$ for Exercise 1.

where $H(i\omega)$ is the transfer function of the system. This is the so-called steady-state response, and is the unique solution that has less than exponential growth as $t \to -\infty$ (see W. Kaplan, pp. 285 et seq.). It is the output that remains after any transients created by the turning on of the input have died out. If $f(t) = 0$ for $t < 0$, then $x(t) = 0$ for $t < 0$, and the output is the solution satisfying zero initial conditions at $t = 0$.

By using the convolution theorem, the output can be expressed as

$$x(t) = h(t) * f(t), \tag{6.91}$$

where $h(t) = \mathscr{F}^{-1}\{H(i\omega)\}$. It is evident that this is nothing other than the impulse response of the system defined in Section 3.5. If the system is characterized by the nth-order differential equation

$$(a_0 D^n + a_1 D^{n-1} + \cdots + a_n)x(t) = f(t),$$

application of a Fourier transform gives

$$X(\omega) = \frac{F(\omega)}{V(i\omega)} = H(i\omega)F(\omega)$$

and when $f(t) = \delta(t)$, implying $F(\omega) = 1$,

$$x(t) = h(t) = \mathscr{F}^{-1}\{H(i\omega)\}.$$

Accordingly, if $f(t) = 0$ for $t < 0$, the output given in (6.90) and (6.91) is identical to that provided by the Laplace-transform technique for a system initially at rest. The main advantages of using Fourier transforms is that the input need not be zero for $t < 0$, but this gain is achieved at the expense of requiring the system to be stable and the input to have *less* than exponential growth at $t = \pm\infty$.

The most fruitful applications of Fourier transforms are based on the interpretation of the transform as a frequency spectrum. With many systems the so-called steady-state response is the only output of interest, and the analysis of the system and its effect on an input is carried out entirely in the frequency domain. The procedure is equivalent to the s-plane analysis discussed in Section 3.7, but with the difference that all of the quantities involved are capable of direct measurement and interpretation. The system is described by its transfer function, which is simply the response to a single-frequency input. This can be measured by injecting a single-frequency oscillation and recording the output in amplitude and phase as the frequency is swept (or stepped) over the band of interest. A simple example of a system is a filter, which is a device for selectively passing some frequencies and attenuating others, and from a knowledge of the spectrum of the input and the filter characteristics, the spectrum of the output is immediately obtained. Quite complicated systems, for example, a frequency modulated (FM) broadcast system, can be analyzed in this manner, and the key to success is often the ability to visualize simultaneously a function and its Fourier transform.

Suggested reading

Sneddon, I. N. *Fourier Transforms*. McGraw-Hill, New York, 1951.

Arfken, G. *Mathematical Methods for Physicists*. 2nd ed. Academic, New York, 1970. A derivation of Fourier transforms from Cauchy's integral formula is given on pp. 313–315.

Papoulis, A. *The Fourier Integral and Its Applications*. McGraw-Hill, New York, 1962.

Kaplan, W. *Operational Methods for Linear Systems*. Addison-Wesley, Reading, MA, 1962. Chapter 5 is concerned with the mathematical aspects of Fourier transforms.

Bracewell, R. N. *The Fourier Integral and Its Applications*. 2nd ed. McGraw-Hill, New York, 1978. A novel and readable presentation of the theory and its applications.

Ziemer, R. E., W. H. Tranter, and D. R. Fannin. *Signals and Systems*. Macmillan, New York, 1983.

Brigham, E. O. *The Fast Fourier Transform*. Prentice-Hall, Englewood Cliffs, NJ, 1974. A detailed description of FFT techniques as well as numerous examples of the Fourier analysis of signals.

CHAPTER 7

Laplace transform revisited

When we last considered the Laplace transform we were only in possession of the direct transform that maps a function of t into a function of the complex variable s. To recover $f(t)$ we had to rely on our ability to recognize the transform or, in practice, to use a table of Laplace transforms, supplemented by certain general properties we had developed. For a unique recovery and, hence, a one-to-one mapping, it is necessary that $f(t) = 0$ for $t < 0$. Mathematically at least, a process of inversion by recognition is unsatisfactory, and since there is an inverse Laplace transform that defines the operation explicitly, we shall now derive it.

7.1 From Fourier to Laplace

In the previous chapter we introduced the Fourier-transform pair consisting of a direct and an inverse transform defined in accordance with Fourier's identity (6.5). The Laplace transform has many similarities to the Fourier and is, in fact, a modified form of a special case, and we can therefore use our knowledge of the inverse Fourier transform to deduce the inverse Laplace. To do so it is necessary to return to the strict definition of a Fourier transform given in Sections 6.2 and 6.3.

The special case referred to is that in which $f(t)$ is zero for $t < 0$. Then

$$f(t) = f_1(t)u(t) \tag{7.1}$$

and if $f_1(t)$ is piecewise-continuous for $0 < t < \infty$ with

$$\int_0^\infty |f_1(t)|^2\, dt < \infty,$$

implying $|f_1(t)| \to 0$ as $t \to \infty$, the Fourier transform of $f(t)$ exists and

$$\mathscr{F}\{f(t)\} = \int_0^\infty f(t)e^{-i\omega t}\,dt = F(\omega) \tag{7.2}$$

with

$$\mathscr{F}^{-1}\{F(\omega)\} = \frac{1}{2\pi}\int_{-\infty}^{\infty} F(\omega)e^{i\omega t}\,d\omega = f(t). \tag{7.3}$$

As indicated by the lower limit of integration in (7.2), $f(t) = 0$ for $t < 0$, and (7.2) and (7.3) represent a *unilateral* or one-sided Fourier-transform pair.

One disadvantage of Fourier transforms, which we have already noted, is that some simple functions such as $u(t)$ do not have transforms in the strict mathematical sense. Nevertheless, it is possible to define a related transform pair associated with the unit-step function. To do this, let

$$f_1(t) = e^{-\sigma t}, \qquad \sigma > 0.$$

The resulting function $f(t)$ given in (7.1) satisfies the conditions for the existence of a Fourier transform, and

$$\int_0^\infty e^{-\sigma t}e^{-i\omega t}\,dt = F(\omega), \tag{7.4}$$

$$\frac{1}{2\pi}\int_{-\infty}^{\infty} F(\omega)e^{i\omega t}\,d\omega = e^{-\sigma t}u(t). \tag{7.5}$$

Clearly,

$$F(\omega) = \frac{1}{i\omega + \sigma}.$$

As long as $\sigma > 0$ the direct transform integral converges absolutely and the pole of $F(\omega)$ at $\omega = i\sigma$ lies in the upper half of the complex ω plane, above the path of integration for the inverse transform. When $\sigma = 0$, $f(t) = u(t)$, but then the integral in (7.4) diverges and, in addition, the inverse transform is meaningless because the pole lies on the path of integration.

We can overcome both difficulties by allowing ω to be complex with a negative imaginary part or, equivalently, by introducing a new variable related to ω. If

$$s = \sigma + i\omega, \tag{7.6}$$

Eq. (7.4) becomes

$$F(\omega) = \int_0^\infty e^{-st}\,dt = F_1(s) \quad \text{(say)}, \tag{7.7}$$

where

$$F_1(s) = \frac{1}{s}.$$

Moreover, from Eq. (7.5) by multiplying through by $e^{\sigma t}$ (which is independent of ω),

$$\frac{1}{2\pi}\int_{-\infty}^{\infty} F(\omega)e^{st}\,d\omega = u(t),$$

and since $d\omega = (1/i)\,ds$, this can be expressed as

$$\frac{1}{2\pi i}\int_{-i\infty+\sigma}^{i\infty+\sigma} F_1(s)e^{st}\,ds = u(t), \tag{7.8}$$

where the integration is along the straight-line path from $s = -i\infty + \sigma$ to $i\infty + \sigma$ in the complex s plane. Equation (7.8) recovers $u(t)$ from a knowledge of the continuous function $F_1(s)$ defined in (7.7).

The essence of the procedure is the inclusion of a factor $e^{-\sigma t}$ in the integrand of the Fourier-transform integral (7.2) to produce convergence, and the rewriting of the direct and inverse (unilateral) Fourier transforms to incorporate the factor into the definitions themselves. We can use this same convergence factor for a more general function $f_1(t)$. If $f_1(t)$ is of exponential order as $t \to \infty$ (see Section 2.3) and σ is such that

$$\int_0^{\infty} e^{-\sigma t}|f_1(t)|\,dt < \infty,$$

implying

$$e^{-\sigma t}|f_1(t)| \to 0 \quad \text{as } t \to \infty,$$

the Fourier transform of $e^{-\sigma t}f(t)$ is

$$\begin{aligned}\mathscr{F}\{e^{-\sigma t}f(t)\} &= \int_0^{\infty} f(t)e^{-\sigma t}e^{-i\omega t}\,dt \\ &= F_1(s)\end{aligned}$$

where

$$F_1(s) = \int_0^{\infty} f_1(t)e^{-st}\,dt,$$

and from Eq. (7.3)

$$\frac{1}{2\pi}\int_{-\infty}^{\infty} F_1(s)e^{i\omega t}\,d\omega = e^{-\sigma t}f_1(t)u(t).$$

On multiplying through by $e^{\sigma t}$ as before and changing the variable of integration from ω to s, we have

$$\frac{1}{2\pi i}\int_{-i\infty+\sigma}^{i\infty+\sigma} F_1(s)e^{st}\,ds = f_1(t)u(t).$$

We recognize $F_1(s)$ as the Laplace transform of $f_1(t)u(t)$; that is,

$$\mathscr{L}\{f_1(t)u(t)\} = \int_0^\infty f_1(t)e^{-st}\,dt = F_1(s) \tag{7.9}$$

and the corresponding inverse transform is

$$\mathscr{L}^{-1}\{F_1(s)\} = \frac{1}{2\pi i}\int_{-i\infty+\sigma}^{i\infty+\sigma} F_1(s)e^{st}\,ds = f_1(t)u(t). \tag{7.10}$$

These constitute the (unilateral) *Laplace-transform pair*, and the definition of the inverse operator $\mathscr{L}^{-1}$ that is applied to a function of s is therefore

$$\mathscr{L}^{-1}\{\cdots\} = \frac{1}{2\pi i}\int_{-i\infty+\sigma}^{i\infty+\sigma}(\cdots)e^{st}\,ds,$$

where σ ($=\operatorname{Re} s$) must be chosen large enough to make the integral in (7.9) converge. By building in the convergence factor $e^{-\sigma t}$, the variety of functions that are Laplace transformable has been increased relative to those for which the unilateral Fourier transform is applicable. We note that if $|f_1(t)| \to 0$ as $t \to \infty$, the Fourier transform (7.2) also exists, and comparison of (7.2) and (7.9) gives

$$\mathscr{F}\{f_1(t)u(t)\} = \mathscr{L}\{f_1(t)u(t)\}|_{s=i\omega},$$

showing that σ can be restricted to nonnegative values (see the remark at the end of Section 2.1).

The inverse Laplace transform provides a representation of a function $f(t)$ that is zero for $t < 0$ and, more importantly, a means of recovering $f(t)$ from a knowledge of its transform. Since the Laplace transform is a (modified) special case of the Fourier, it follows that if $f(t)$ is piecewise-smooth and σ is such that the direct transform integral converges absolutely, $F_1(s)$ is a continuous function of s. The inverse transform integral then converges and the value of the integral is $f(t)$ wherever $f(t)$ is continuous, and is the average of the right- and left-hand limits at every point of jump discontinuity in $f(t)$. Except possibly at such points, a piecewise-smooth function is uniquely determined by its Laplace transform. As a matter of fact, this is true also if $f(t)$ is only piecewise-continuous (see, for example, W. Kaplan, pp. 317–18).

7.2 Evaluating the inverse transform

If $F(s)$ is the Laplace transform of $f(t)$, then

$$f(t) = \mathscr{L}^{-1}\{F(s)\} = \frac{1}{2\pi i}\int_{-i\infty+\sigma}^{i\infty+\sigma} F(s)e^{st}\,ds, \tag{7.11}$$

where, for simplicity, we have omitted the suffix unity. In contrast to the inverse Fourier transform, the inverse Laplace transform is an integral in the complex plane, which is the reason it was not discussed in Chapter 2. Evaluation by direct integration along the path is almost always a complicated task, but for a wide class of functions $F(s)$ the evaluation can be carried out by closing the contour to the right or to the left and using the residue method. Fortunately, these functions are the ones of most interest in systems theory.

Let $F(s)$ be a proper rational function of s, or the product of such a function with an exponential of the form e^{-as}, where a is real with $a \geq 0$. The limitation on a is a consequence of the t-plane shift property (2.14) of a Laplace transform, and the fact that σ must be chosen to ensure convergence of the direct transform integral (7.9) also places the path of integration for the inverse transform to the right of all singularities of $F(s)$. In a half-plane to the right of the path and including the path itself, $F(s)$ is therefore analytic as a function of s. Since the rational function is proper, the conditions for path closure (see Section 5.7) are satisfied in either the half-plane to the right (RHP) or the half-plane to the left (LHP) depending on the value of t, and because of the restriction on σ, closure to the right gives zero. On closing to the left we find

$$f(t) = 2\pi i \sum_n \operatorname{Res}\left\{\frac{1}{2\pi i}F(s)e^{st}, s_n\right\}$$

and thus

$$f(t) = \sum_n \operatorname{Res}\{F(s)e^{st}, s_n\},$$

where the s_n are the poles of $F(s)$. The procedure is simple, as evident from the following examples.

Example 1. Determine $\mathscr{L}^{-1}[e^{-s}/(s^2-4)]$.

From the definition (7.11) we have

$$f(t) = \frac{1}{2\pi i}\int_{-i\infty+\sigma}^{i\infty+\sigma} \frac{e^{-s}}{s^2-4}e^{st}\,ds$$

$$= \int_{-i\infty+\sigma}^{i\infty+\sigma} g(z)\,dz$$

with

$$g(z) = \frac{1}{2\pi i} \frac{e^{z(t-1)}}{(z-2)(z+2)}$$

and $\sigma > 2$. Since

$$e^{z(t-1)} = e^{x(t-1)} e^{iy(t-1)},$$

the requirement that $x(t-1)$ be negative in the half-plane of closure shows that

$$\text{if} \quad t - 1 < 0, \quad \text{then} \quad x > 0\,(\text{RHP}),$$

whereas

$$\text{if} \quad t - 1 > 0, \quad \text{then} \quad x < 0\,(\text{LHP}).$$

For $t - 1 < 0$ closure to the right gives zero since there are no singularities there, but for $t - 1 > 0$ closure to the left gives

$$\begin{aligned} f(t) &= 2\pi i[\text{Res}\{g(s), 2\} + \text{Res}\{g(s), -2\}] \\ &= 2\pi i \frac{1}{2\pi i}\left\{\frac{e^{2(t-1)}}{4} - \frac{e^{-2(t-1)}}{4}\right\} \\ &= \frac{1}{4}\{e^{2(t-1)} - e^{-2(t-1)}\}. \end{aligned}$$

Hence, for all t,

$$f(t) = \tfrac{1}{2}\sinh 2(t-1)u(t-1)$$

and we observe that $f(t)$ is continuous.

Example 2. Determine $\mathscr{L}^{-1}\{1/s\}$.

We have

$$f(t) = \frac{1}{2\pi i}\int_{-i\infty+\sigma}^{i\infty+\sigma} \frac{e^{st}}{s}\, ds$$

with $\sigma > 0$. For $t < 0$ closure of the path to the right gives zero, whereas for $t > 0$ closure to the left gives

$$f(t) = \text{Res}\left\{\frac{e^{st}}{s}, 0\right\} = 1.$$

Hence

$$f(t) = u(t),$$

which is discontinuous at $t = 0$. When $t = 0$ the conditions for path closure are violated, but the inverse transform can still be evaluated by direct integration along the path (see Exercise 10 following). Thus, an inverse Laplace transform (like the Fourier) defines a function even at a point of jump discontinuity and, as expected, $f(0) = \frac{1}{2}$.

In the first example we included some of the details of the calculation to show how the half-plane for closure is specified by the range of t, leading to the determination of $f(t)$ for (almost) all t, $-\infty < t < \infty$. In practice such details can be omitted, and if

$$F(s) = G(s)e^{-as},$$

where $G(s)$ is a proper rational function, it follows immediately that

$$f(t) = \sum_n \text{Res}\{G(s)e^{s(t-a)}, s_n\}u(t-a),$$

where the s_n are the poles of $G(s)$. When $F(s)$ consists of two or more terms with different exponentials, each must be considered separately.

For the path of integration in Eq. (7.11), only the least value of σ corresponding to the boundary of the half-plane of analyticity of $F(s)$ is specified by the requirement for absolute convergence of the direct transform integral. When evaluating $\mathscr{L}^{-1}\{F(s)\}$ any larger value can be chosen, and it is easy to show (see, for example, p. 159 et seq. of the book by R. V. Churchill) that $f(t)$ is independent of the choice. The reader may also be concerned at our use of Jordan's lemma when the path of integration does not coincide with a coordinate axis. After all, when $\sigma \neq 0$ the so-called left half-plane includes part of the region to the right of the imaginary axis, but a simple change of variable eliminates the problem. If, in the first example, we choose $\sigma = 3$, the substitution $s' = s - 3$ gives

$$f(t) = \frac{1}{2\pi i}\int_{-i\infty}^{i\infty} \frac{1}{(s'+1)(s'+4)} e^{s'(t-1)}\,ds'\,e^{3(t-1)},$$

and the applicability of Jordan's lemma is now obvious. It can be verified that the result obtained is the same as before.

If the rational function is improper, the inverse transform integral does not converge in a strict mathematical sense, and the conditions for path closure are no longer satisfied. In the context of the system problems discussed in Chapter 3, this situation arises when the function $F(s)$ is attributable in part to generalized functions. By dividing the denominator into the numerator, $F(s)$ can be expressed as a polynomial in s plus a proper rational function. The latter can be treated using the method described and each term in the polynomial can be interpreted using Eq. (3.16).

Exercises

Using the method of residues, find the inverse Laplace transforms of the following functions.

*1. $\dfrac{1}{s^2 - 1}$.

2. $\dfrac{1}{(s+1)^2}$.

3. $\dfrac{s^2}{(s^2 - 1)^2}$.

4. $\dfrac{s}{(s+1)(s+2)}$.

*5. $\dfrac{e^{-2s}}{(s-a)^4}$.

6. $\dfrac{1}{s^2 + 2s + 5}$.

7. $\dfrac{1 - e^{-s}}{(s+1)^3}$.

8. $\dfrac{e^{-s}}{s^2 + 1}$.

9. $\dfrac{1}{s^3 + 1}$.

10. By direct integration along the path, show that $\mathscr{L}^{-1}\{1/s\} = \frac{1}{2}$ when $t = 0$.

7.3 Why Laplace?

The Fourier and Laplace transforms have many features in common, and a variety of tasks can be accomplished with either. Both of them simplify functions, converting, for example, exponential and other functions into algebraic quantities, and handling with relative impunity functions that are only piecewise-continuous. They also simplify operations. Convolution is converted to a product, and differentiation and integration are reduced to multiplication and division, respectively. In this respect the transformations are analogous to those performed by a logarithm, but whereas a logarithm deals with numbers, Fourier and Laplace transforms deal with functions.

The main advantage of the Fourier transform is its physical interpretability. The inverse transform or Fourier integral is a representation of $f(t)$ in terms of undamped sinusoidal oscillations, and the transform has an immediate interpretation as a frequency spectrum. Because of this, in calculations involving Fourier transforms it is often possible to give a meaning to intermediate stages of the analysis, and to retain a physical grasp of the problem throughout. In contrast, the interpretation of a Laplace transform is less obvious, and the fact that the inverse transform is a representation in terms of exponentials does not convey the same amount of physical information. Nevertheless, the Laplace transform has its own advantages, which make it indispensable for treating systems problems involving electrical or other transients. In the solution of a differential equation it incorporates the initial conditions automatically and eliminates the need for a separate, time-consuming operation to insert the initial values. Functions that grow exponentially as well as those that decay are handled with equal facility, as is the unit-step function, representing the basic turn-on operation. Last but not least, the Laplace transform of a real function of t is a real function of s, and this is certainly an advantage in any calculation. In spite of our attempts to demonstrate the power and the glory of complex variables, it is still true that real quantities are more simple than complex.

Suggested reading

Kaplan, W. *Operational Methods for Linear Systems.* Addison-Wesley, Reading, MA, 1962.

Churchill, R. V. *Modern Operational Mathematics in Engineering.* McGraw-Hill, New York, 1944.

Answers to selected exercises

These exercises were denoted by an asterisk in the text.

Page 4,	Exercise 1:	$-21-20i$.
	Exercise 8:	1.
Page 10,	Exercise 1:	$-1-i$.
	Exercise 7:	$3-2i, -3+2i$.
	Exercise 11:	those lying on a circle of radius 2 centered at $x=2, y=0$.
Page 17,	Exercise 1:	$2/s^3$ with $\sigma > 0$.
	Exercise 5:	$s/(s^2-\alpha^2)$ with $\sigma > \alpha$.
Page 21,	Exercise 1:	$e^{-t^2}\{u(t)-u(t-2)\}$.
	Exercise 3:	$1/(s+1)^2$ with $\sigma \geq 0$.
Page 30,	Exercise 1:	$24/(s-1)^5$.
	Exercise 5:	$2e^{-s}/(s^2+4)$.
	Exercise 13:	$\frac{1}{s}\frac{1-e^{-s}}{1+e^{-s}}$.
Page 42,	Exercise 1:	$-\frac{2}{s+1}+\frac{4}{s+2}$.
	Exercise 7:	$\frac{2}{s^2}-\frac{3}{s}+\frac{4}{s+1}-\frac{1}{s+2}$.
Page 44,	Exercise 1:	$2(\cosh t)u(t) \to 2(\sinh t)u(t)$.
	Exercise 5:	$(1-\cos t)u(t)$.
	Exercise 11:	$\{e^{2(t-2)}-e^{t-2}\}u(t-2)$.
Page 48,	Exercise 1:	$x(t)=(\sin 2t-\cos 2t+\cos t)u(t)$ and $x(t) = \sin 2t - \cos 2t + \cos t$ for $-\infty < t < \infty$.
	Exercise 5:	$x(t)=(3e^{-t}-\cos t+\sin t)u(t)$ and $x(t) = 2e^{-t}+(e^{-t}-\cos t+\sin t)u(t)$ for $-\infty < t < \infty$.

Exercise 11: $x(t) = (2e^{-t} - e^{-3t} + 2\sin t)u(t)$.

Page 64, Exercise 2(i): $u(t) + \delta(t), \delta(t) + \delta'(t)$.

Exercise 3(i): 1.

Exercise 5: $e^{-t}u(t) + 4\delta(t) + \delta'(t)$.

Page 74, Exercise 2: $\{(2t + 1)e^{-t} - (4t + 1)e^{-3t}\}u(t)$.

Exercise 9: $e^{-t}(\cos t)u(t)$.

Page 81, Exercise 1: $\frac{1}{3}(\sin t - \frac{1}{2}\sin 2t)u(t)$.

Page 88, Exercise 1: $x_1(t) = (3e^{4t} + 5e^{-t})u(t)$,
$x_2(t) = (-2e^{4t} + 5e^{-t})u(t)$.

Page 95, Exercise 1(i): stable.

Exercise 4(i): $(2 - 2e^{-t} - te^{-t})u(t)$.

Page 108, Exercise 1:
$$\frac{1}{4} + \sum_{n=1}^{\infty} \frac{1}{n\pi}\left\{\sin\frac{n\pi}{2}\cos\frac{n\pi t}{2} + \left(1 - \cos\frac{n\pi}{2}\right)\sin\frac{n\pi t}{2}\right\}$$

Exercise 7:
$$\frac{1}{\pi} + \sum_{n=1}^{\infty} \frac{2/\pi}{4n^2 - 1} \times \left\{(-1)^{n-1}\cos 2nt + 2n\sin 2nt\right\}.$$

Page 127, Exercise 1:
$$-\frac{\pi}{4} + \sum_{n=1}^{\infty} \frac{1}{n}\left\{-\frac{1}{n\pi}(1 - (-1)^n)\cos nt + (1 - 2(-1)^n)\sin nt\right\}.$$

Exercise 5:
$$\sum_{n=1}^{\infty}\left(\frac{2}{n\pi}\right)^2\left(\frac{n\pi}{2} - \sin\frac{n\pi}{2}\right)\sin\frac{n\pi t}{2}.$$

Exercise 10:
$$\frac{1}{\pi} + \frac{4}{3\pi}\cos\frac{t}{2} + \sum_{n=3}^{\infty}\left(-\frac{4}{\pi}\right)\frac{1 + \cos(n\pi/2)}{n^2 - 4}\cos\frac{nt}{2}.$$

Page 132, Exercise 1:
$$\sum_{n=-\infty}^{\infty} \frac{1 - 1/e}{1 + 2\pi in} e^{2\pi int}.$$

Page 141, Exercise 1(i): the straight lines $v = -u - 2$ and $v = u + 2$.

Exercise 8: $e^{-i\pi/4}$ and $-i$.

Page 152, Exercise 1(i): $-24z(1 - 3z^2)^3$.

Exercise 4(i): $v(x, y) = -e^x\cos y - x + c$,
$f(z) = -i(e^z + z + c)$.

Page 160, Exercise 1(i): $2i$.

Exercise 6(i): 0.

Page 171, Exercise 1(i): $2\pi i$.

Exercise 4(i): π.

Page 183, Exercise 1(i): $1 + \frac{z^2}{2!} + \frac{z^4}{4!} + \cdots$

$= \sum_{n=0}^{\infty} \frac{z^{2n}}{(2n)!} \quad (|z| < \infty).$

Exercise 4(i): $\frac{1}{z} + \frac{1}{z^3} + \frac{1}{z^5} + \cdots \quad (|z| > 1).$

Page 190, Exercise 1(i): $-1.$

Exercise 3(i): $i\pi/2.$

Exercise 6(i): $3\pi/4.$

Page 203, Exercise 1: $\pi/2.$

Exercise 9: $\frac{\pi}{ab(a+b)}.$

Exercise 19: $\frac{\pi}{2} \csc \frac{a\pi}{2}.$

Page 219, Exercise 1: $\frac{1/\pi}{1+t^2}.$

Exercise 6: $\frac{1}{i\omega} + \frac{1}{\omega^2 t_0}(1 - e^{-i\omega t_0}).$

Exercise 11: $f(t) = \frac{i}{\pi}\int_{-\infty}^{\infty} \frac{1}{\omega}\left(1 - \frac{\sin\omega}{\omega}\right)e^{i\omega t}\,d\omega.$

Page 230, Exercise 1: $\frac{\sin[1/4(\omega - \omega_0)]}{\omega - \omega_0} + \frac{\sin[1/4(\omega + \omega_0)]}{\omega + \omega_0}.$

Exercise 7: $\frac{1}{2}(e^{-t} - e^{-3t})u(t).$

Page 240, Exercise 13: $\frac{1}{2}e^{-2|t|}.$

Page 263, Exercise 1: $(\sinh t)u(t).$

Exercise 5: $\frac{1}{6}(t-2)^3 e^{a(t-2)}u(t-2).$

Index

In the following list, only the more important citations are given, with the primary ones in italic type where appropriate.